SCIENCE CHUNKS

Year B Guide

Science Chunks - Year B

First Edition 2021

Digital Edition

Email: support@elementalscience.com

Copyright Policy

Year B Overview

Welcome to Year B of the Science Chunks series! The seven units that make up this year will cover the human body, weather, stars, matter, acids, bases, and energy, plus a short unit on Louis Pasteur. You will learn about these subjects by:

- **Reading** the assigned encyclopedia pages with your students to learn about the lesson's topic.
- **Writing** down what the students have learned in a way that is appropriate for their skills using either the lapbook templates or the notebook sheets.
- **Doing** one or more of the lesson's coordinating STEAM activities.

What do we need?

In addition to the lesson plans found in the seven units included in Year B, you will need the following materials:

1. **Required Books** – See a complete list on p. 8 of this guide or visit the Year B resource page for links to the books:
   - https://elementalscience.com/blogs/resources/sc-b
2. **Student materials** – Unlike the single Science Chunks units, this printed Year B guide does not inlcude the student materials. You will need to purchase one (or both) of the following:
   - *Science Chunks Year B Student Lapbook Templates (LT)*
   - *Science Chunks Year B Student Notebook (NB)*
3. **Supplies** – You will need various supplies for the lapbook or the notebook, plus the coordinating hands-on science activities. You can find a full list of what you need for the year on pp. 11-12.

How do we organize the year?

Generally, we recommend completing one lesson per week. Please feel free to organize the order of the units in a way that fits your goals. That said, here are two suggestions:

Option 1 (By Discipline)
- [ ] Human Body Unit (p. 13)
- [ ] Weather Unit (p. 37)
- [ ] Stars Unit (p. 51)
- [ ] Matter Unit (p. 65)
- [ ] Acids and Bases Unit (p. 75)
- [ ] Pasteur Unit (p. 83)
- [ ] Energy Unit (p. 87)

Option 2 (By Season - Starting in the Fall)
- [ ] Stars Unit (p. 51)
- [ ] Matter Unit (p. 65)
- [ ] Human Body Unit (p. 13)
- [ ] Acids and Bases Unit (p. 75)
- [ ] Pasteur Unit (p. 83)
- [ ] Energy Unit (p. 87)
- [ ] Weather Unit (p. 37)

If you have any questions or comments as you work through these materials, please don't hesitate to let us know by emailing support@elementalscience.com.

Table of Contents

A Peek Inside a Science Chunks Unit

1. Lesson Topic

Focus on one main idea throughout the week. You will learn about these ideas by reading from visually appealing encyclopedias, recording what the students learned, and doing coordinating hands-on science activities.

2. Information Assignments

Find two reading options—one for younger students, one for older students, plus optional library books.

3. Notebooking Assignments

Record what your students have learned with either a lapbook or a notebook. The directions for these options are included for your convenience in this section along with the vocabulary the lesson will cover.

4. Hands-on Science Assignments

Get the directions for coordinating hands-on science activities that relate to the week's topic.

5. Lesson To-Do Lists

See what is essential for you to do each week and what is optional. You can check these off as you work through the lesson so that you will know when you are ready to move on to the next one.

6. Lapbook Templates

Get all the information you need to create a lapbook on the subject. *(Year bundle sold separately.)*

7. Notebook Templates

Have all the sheets you need to create a notebook on the subject, including a glossary for the vocabulary terms. *(Year bundle sold separately.)*

In the appendix you will find a blank activity sheet, a blank lab report sheet, and any necessary templates.

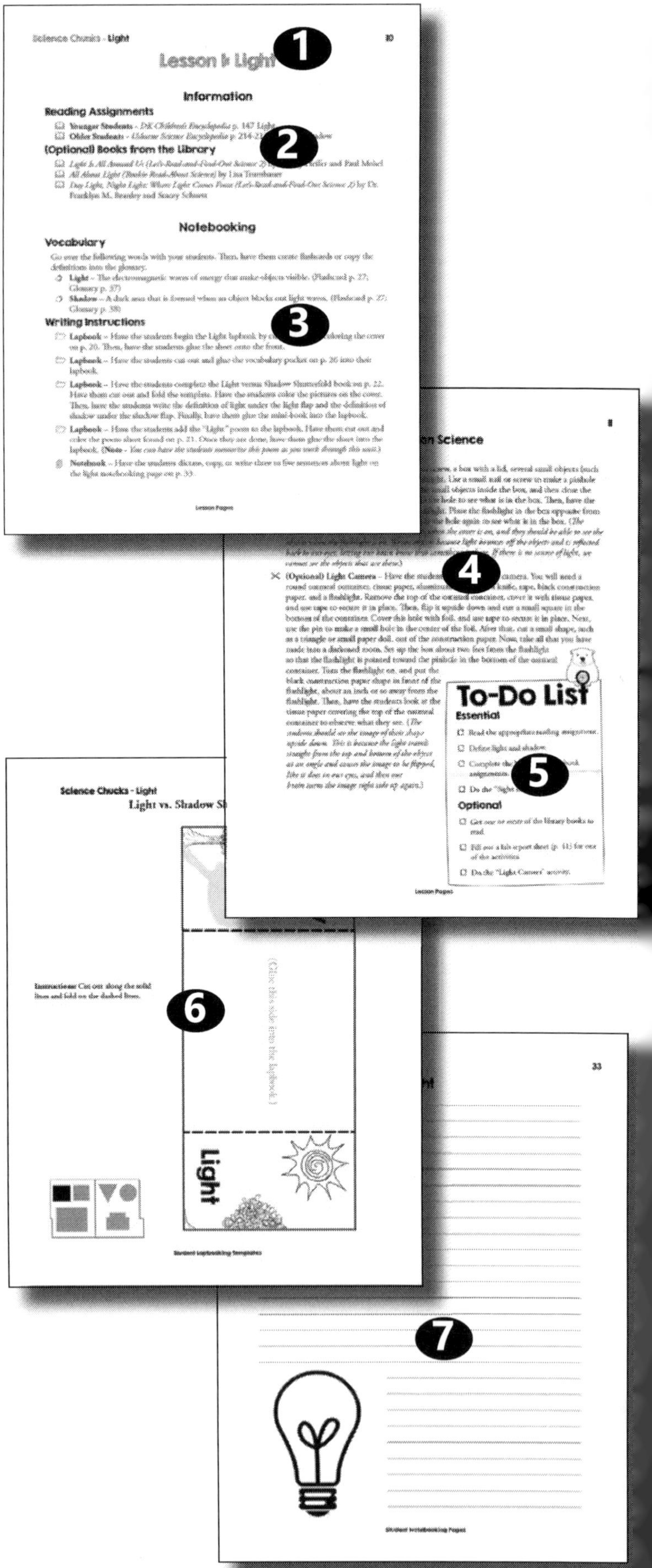

Introduction

Science Chunks - Year B is a unique and versatile unit study that leads you through a survey of the human body, weather, stars, matter, acids, bases, and energy, plus a short unit on Louis Pasteur. It is designed to be a gentle approach to homeschool science based on the Unit Study method suggested in *Success in Science: A Manual for Excellence in Science Education* by Bradley and Paige Hudson. This study can be used as a stand-alone unit for elementary science.

What Is Included in Each Unit

Science Chunks - Year B includes seven units, each with the three keys to teaching science. For a lesson you will be doing the following:

- ✓ Listening to (or reading) **scientific information** from visually appealing encyclopedias
- ✓ Dictating (or writing down) what the students have learned and seen using **lapbooking or notebooking**
- ✓ Watching (and doing) **hands-on science** through a variety of science activities

Here is how this works.

Section 1 - Information

The elementary student is an empty bucket waiting to be filled with information, and science-oriented books are a wonderful way to do that. These books can include age-appropriate children's science encyclopedias, living books for science, and/or children's nonfiction science books.

In this program, the reading assignments and additional books scheduled in the lesson fulfill this component. The reading assignments are broken for you into two levels: younger students (1st to 3rd grade) and older students (4th to 6th grade).

Our idea is that you will read these selections with your students, pausing to ask questions or discussing the information once you are done reading.

Section 2 - Notebooking

The purpose of the notebooking component for elementary science education is to verify that the students have placed at least one piece of information into their knowledge bucket. You can use notebooking sheets, lapbooks, and/or vocabulary words to fulfill this requirement.

In this program, we have created two writing options, a lapbook and a notebook, for you to use with your students. In *Science Chunks Year B Student Lapbook Templates*, you will find all of the templates and pictures you will need to complete seven different lapbooks for the year. In *Science Chunks Year B Student Notebook*, you will find all the pages you need to create a simple notebook documenting what the students learned this year, including notebooking sheets and a glossary for each unit.

Section 3 - Hands-on Science

Scientific demonstrations and observations are meant to spark students' enthusiasm for learning science, to work on their observation skills, and to demonstration the principles of science for them. This component of elementary science education can contain scientific demonstrations, hands-on projects, and/or nature studies.

In this program, the coordinating activities at the end of each lesson fulfill this section of elementary science instruction. If you would like to record what you have done, you can use one of the templates in the appendix pp. 100-101.

What You Need in Addition to This Guide

Books Scheduled

The following books are what we used to plan the reading assignments for this year:

- **Younger students (1st to 3rd grade)** - *DK First Human Body Encyclopedia, Chemistry: Getting a Big Reaction (Basher Science), The DK Children's Encyclopedia*
- **Older students (4th to 6th grade)** - *Usborne Science Encyclopedia, Kingfisher Science Encyclopedia*
- **All students** - *Glow in the Dark Constellations, Pasteur's Fight Against Microbes* by Beverly Birch

However, you could certainly use the encyclopedias you already have on hand or books from the library.

Unlike the single Science Chunks units, this printed Year B guide does not inlcude the student materials. You will need to purchase one (or both) of the following:

- *Science Chunks Year B Lapbook Templates (LT)*
- *Science Chunks Year B Student Notebook (NB)*

In addtion to the books and the student materials, you will need also simple craft supplies and other science materials—see a complete list of essential items on pp. 11-12.

How This Unit Works

We have included a to-do list with each lesson to give you an idea of what is essential and what is optional. There are several ways you can schedule this unit. Here is a quick look at a few of the options.

Possible Schedules for Your Week

- **One Day** – You can set aside about an hour to an hour and a half each week to complete all the essential tasks in one day.

- **Two Days** – You can set aside about 30 to 40 minutes twice a week to complete all the essential tasks, plus a few more, in two days. On the first day, you can complete the reading assignments and either the lapbook or notebook assignments. On the second day, you can complete the coordinating activity and the vocabulary assignments as well as read any library books.
- **Three Days** – You can set aside about 30 minutes three times a week to complete all the essential tasks, plus a few more, in three days. On the first day, you can complete the reading assignments and either the lapbook or notebook assignments. On the second day, you can complete the coordinating activity and write a lab report using one of the templates. On the third day, you can do the vocabulary assignments as well as read any library books.
- **Four Days** – You can set aside about 20 to 30 minutes four times a week to complete all the essential tasks, plus a few more, in four days. On the first day, you can complete the reading assignments and either the lapbook or notebook assignments. On the second day, you can complete the coordinating activity and write a lab report. On the third day, you can do the vocabulary assignments as well as read any library books. On the fourth day, you can do the optional coordinating activity as well as read any library books.

Generally, we recommend completing one lesson per week, but please feel free to organize the units in the order that fits your goals.

Final Thoughts

Quick Links

View all the links mentioned in this guide in one place and get a digital copy of the appendix templates, glossary, and review sheets by visiting the following page:

https://elementalscience.com/blogs/resources/sc-b

Read Further

If you would like to read more about the philosophy behind the Science Chunks series, check out *Success in Science: A Manual for Excellence in Science Education* and the following articles from our website.

- **The Three Keys to Teaching Science** – This article shares the three keys to teaching science, including a free session that walks you through what each key can look like.
  - https://elementalscience.com/blogs/news/3-keys
- **The Basics of Notebooking** – This article details the basic components of notebooking along with how a few suggestions on what notebooking can look like.
  - https://elementalscience.com/blogs/news/what-is-notebooking
- **Lapbooking versus Notebooking** – This article takes a look at the differences between lapbooking and notebooking.
  - https://elementalscience.com/blogs/news/lapbook-or-notebook

- **Scientific Demonstrations versus Experiments** – This article explains the difference between scientific demonstrations and experiments along with when and how to employ these methods.
  - https://elementalscience.com/blogs/news/89905795-scientific-demonstrations-or-experiments

Last Words

As the author and publisher of this curriculum, I encourage you to contact me with any questions or problems that you might have concerning *Science Chunks - Year B* by emailing us at support@elementalscience.com. I, or a member of our team, will be more than happy to answer them as soon as we can. I hope that you will enjoy creating memories this year!

~ Paige Hudson

Materials List

Human Body Unit Materials

You will need the following materials to complete the essential coordinating activities:

- ✂ **Lesson 1:** Margarine container, Jell-O, Grape, Options for organelles
- ✂ **Lesson 2:** Paper, Tape, Several students or family members
- ✂ **Lesson 3:** 5 to 6 Spools of thread, 2 Cardboard discs, String
- ✂ **Lesson 4:** Brad, Card stock, Scissors, Hole punch, 2 - 1x2 boards, Hinge, Screwdriver, Screws
- ✂ **Lesson 5:** Book, See-through barrier like a window or a wire screen, Cotton ball
- ✂ **Lesson 6:** Different types of food, Paper plate, Cloth
- ✂ **Lesson 7:** Paper towel tube
- ✂ **Lesson 8:** *No supplies needed*
- ✂ **Lesson 9:** Large Ziploc bag, Bread, Coke
- ✂ **Lesson 10:** Balloon, Funnel, Water
- ✂ **Lesson 11:** Sterile cotton swabs, Petri dishes, Agar

Weather Unit Materials

You will need the following materials to complete the essential coordinating activities:

- ✂ **Lesson 1:** Plastic water bottle, Duct tape, Permanent marking pen, Small marbles or rocks, Ruler
- ✂ **Lesson 2:** Lump of air dry clay, Several brown pipe cleaners, Felt sheets of either green, red, orange, or yellow
- ✂ **Lesson 3:** Paper or cotton ball, Straw for each of the students
- ✂ **Lesson 4:** Glass jar or cup, Plastic wrap, Rubber band, Ice cubes
- ✂ **Lesson 5:** Clear glass, Shaving cream, Blue food coloring, Warm water
- ✂ **Lesson 6:** Shaving cream, Cornstarch

Stars Unit Materials

You will need the following materials to complete the essential coordinating activities:

- ✂ **Lesson 1:** Black construction paper, Toothpick, Tape, Flashlight, Large, clear bowl, Water
- ✂ **Lesson 2:** Foil, Toilet paper tube, Pin, Small flashlight, Rubber band, Sharpie marker
- ✂ **Lesson 3:** Flashlight planetarium (*from previous lesson*), Foil, Pin, Sharpie marker
- ✂ **Lesson 4:** Flashlight planetarium (*from previous lesson*), Foil, Pin, Sharpie marker
- ✂ **Lesson 5:** Flashlight planetarium (*from previous lesson*), Foil, Pin, Sharpie marker
- ✂ **Lesson 6:** Flashlight planetarium (*from previous lesson*), Foil, Pin, Sharpie marker

Matter Unit Materials

You will need the following materials to complete the essential coordinating activities:

- **Lesson 1:** 3 Balloons, Water, Ice
- **Lesson 2:** Orange Juice, Cup
- **Lesson 3:** 3 Toothpicks, Dish Soap, Bowl, Water
- **Lesson 4:** 2-Liter soda bottle, Quarter, Water

Acids and Bases Unit Materials

You will need the following materials to complete the essential coordinating activities:

- **Lesson 1:** Head of red cabbage, Knife, Pot, Water, Strainer, Large plastic container, 2 Clear glasses, ammonia, White vinegar
- **Lesson 2:** Head of cabbage, Water, Pot, Variety of liquids from your kitchen (ex. lemon juice, baking soda, powder, or detergent), Several cups
- **Lesson 3:** Vinegar, Baking soda, Water, Cabbage juice, Coffee filter, 2 Clear cups, Eyedropper

Pasteur Unit Materials

You will need the following materials to complete the essential coordinating activities:

- **Lesson 1:** A packet of yeast, Warm water, Sugar, Clear container

Energy Unit Materials

You will need the following materials to complete the essential coordinating activities:

- **Lesson 1:** Several people, Rubber band for each person, Measuring tape
- **Lesson 2:** Scissors, Plastic wrap, Aluminum foil, Tape, Old pizza box, Black construction paper, Glass or metal pie plate, Marshmallows or piece of buttered toast
- **Lesson 3:** 32 Pieces bite-sized food such as M&M's, raisins, or cereal puffs, Timer
- **Lesson 4:** 3 Bowls, hot-to-the-touch (but not burning) water, Cold water, Room temperature water
- **Lesson 5:** Sunny window sill, Bowl, Cool water

Abbreviations Key

LT = *Science Chunks Year B Student Lapbook Templates*

NB = *Science Chunks Year B Student Notebook*

SCIENCE
CHUNKS

Human Body Unit

Lesson 1: Basic Building Blocks

Information

Reading Assignments

- **Younger Students** - *DK First Human Body Encyclopedia* pp. 6-7 What makes you you?, pp. 8-9 Building Blocks
- **Older Students** - *Kingfisher Science Encyclopedia* pp. 98-99 Body Organization

(Optional) Books from the Library

- *Have a Nice DNA (Enjoy Your Cells, 3)* by Fran Balkwill and Mic Rolph
- *Cells: An Owner's Handbook* by Carolyn Fisher

Notebooking

Vocabulary

Go over the following word with your students. Then, have them create a flashcard or copy the definition into the glossary.

- **Cells** – A tiny, living units that is the building blocks of all living things. (Flashcard LT p. 20; Glossary NB p. 57)

Writing Instructions

- **Lapbook** – Have the students cut out and glue the vocabulary pocket on LT p. 20 into their lapbook.
- **Lapbook** – Have the students cut out the "Building Blocks" tab-book on LT p. 8 and color the pictures. Have the students tell you what they have learned about cells and DNA. Write it on the respective tabs for them. Then, have them staple the pages together and glue the mini-book into the lapbook.
- **Notebook** – Have the students dictate, copy, or write one to four sentences for cells and DNA on the building blocks notebooking page on NB p. 6.

Hands-on Science

Coordinating Activities

- **Cell Model** – Have the students make a Jell-O replica of a cell. Use a margarine container for your cell membrane, Jell-O for cytoplasm. Prepare the Jell-O according to the package directions. Fill your container three-quarters of the way with the mixture and place it in the fridge until it is soft set (about 30 minutes). Then, insert a grape in the center for the nucleus of the cell and use your imagination for materials for the remaining organelles.

- **(Optional) DNA Ladder** – Have the students build their own DNA ladders. You will need four different colors of LEGO® brick blocks (at least six of each). You can get directions for this project at the following website:
  - https://elementalscience.com/blogs/science-activities/simple-lego-dna-tower

To-Do List

Essential

- ☐ Read the appropriate reading assignment.
- ☐ Define cells.
- ☐ Complete the lapbook or notebook assignments.
- ☐ Do the "Cell Model" activity.

Optional

- ☐ Get one or more of the library books to read.
- ☐ Fill out a lab report sheet (p. 101) for one of the activities.
- ☐ Do the "DNA Ladder" activity.

Lesson 2: Skin and Hair

Information

Reading Assignments

- **Younger Students** - *DK First Human Body Encyclopedia* pp. 68-69 All Wrapped Up, pp. 72-73 Fairly Hairy
- **Older Students** - *Kingfisher Science Encyclopedia* pp. 100-101 Skin, Hair, and Nails

(Optional) Books from the Library

- *Body Works - Skin and Hair* by Katherine Goode
- *Your Skin and Mine: Revised Edition (Let's-Read-and-Find... Science 2)* by Paul Showers
- *How Do Hair and Nails Grow?* by Thomas Young
- *Inside the Skin, Hair, & Nails (Super Simple Body)* by Karin M D Halvorson

Notebooking

Vocabulary

There are no vocabulary words for this lesson.

Writing Instructions

- **Lapbook** – Have the students cut out the "Skin and Hair" tab-book on LT p. 9 and color the pictures. Have the students tell you what they have learned about skin and hair. Write it on the respective tabs for them. Then, have them staple the pages together and glue the mini-book into the lapbook.
- **Notebook** – Have the students dictate, copy, or write one to four sentences for skin and hair on the skin and hair notebooking page on NB p. 7.

Hands-on Science

Coordinating Activities

- **Hair Compare** – Have the students compare their hair with the hair of others in their family or group. You will need paper, tape, and several students or family members for this activity. Pull a hair from each person and tape each of them on a sheet of paper. Mark each one with a number and record the person's name with that number on a separate sheet of paper. Then, have each of the students guess at whose hair belongs to who.
- **(Optional) Human Body Project** – Each week, the students will paste the parts of the body that you are learning about onto the body outline. Add each part at the end of the week while

doing an oral review of what you have learned by asking, "Do you remember what this part does?". This project will continue throughout the unit. This week, have the students add hair to the body outline on pp. 105-106.

To-Do List

Essential

- ☐ Read the appropriate reading assignment.
- ☐ Complete the lapbook or notebook assignments.
- ☐ Do the "Hair Compare" activity.

Optional

- ☐ Get one or more of the library books to read.
- ☐ Fill out a lab report sheet for one of the activities.
- ☐ Do the "Human Body Project" activity.

Lesson 3: Skeletal System

Information

Reading Assignments

- **Younger Students** - *DK First Human Body Encyclopedia* pp. 12-13 Skeleton, pp. 16-17 Bendable Backbone, pp. 14-15 Head Case, pp. 20-21 Bones & Cartilage
- **Older Students** - *Kingfisher Science Encyclopedia* pp. 102-103 The Skeleton, p. 104 Bones

(Optional) Books from the Library

- *The Skeleton Inside You (Let's-Read-and-Find... Science 2)* by Philip Balestrino
- *Bones: Skeletons and How They Work* by Steve Jenkins
- *A Book about Your Skeleton (Hello Reader!)* by Ruth Belov Gross

Notebooking

Vocabulary

Go over the following word with your students. Then, have them create a flashcard or copy the definition into the glossary.

- **Skeleton** – The framework of 206 bones that supports your body; it allows you to move and protects certain organs. (Flashcard LT p. 21; Glossary NB p. 65)

Writing Instructions

- **Lapbook** – Have the students cut out and color the "The Systems of the Human Body" poem on LT p. 10. Once they have finished, have them glue the poem into the lapbook.
- **Lapbook** – Have the students cut out the "Skeletal System" tab-book on LT p. 11 and color the cover. Then, have them tell you the purpose (or job) of the skeletal system and record it on the purpose page. After that, have the students tell you what they have learned about bones and the skull. Write it for them on the respective pages. Finally, staple the pages together and glue the tab-book into the lapbook.
- **Notebook** – Have the students dictate, copy, or write one to four sentences for the skeleton, skull, and bones on the skeletal system notebooking page on NB p. 8.

Hands-on Science

Coordinating Activities

- **Backbone Model** – Have the students make a model of the backbone. You will need 5 to 6 spools of thread, 2 cardboard discs and some string for this project. Tie one end of one of the cardboard discs to the string, then thread the string through the center of the spools of thread. Now tie the remaining cardboard discs to the other end of the string, close to the top of the last spool of thread. You now have a model of the spine that will twist and bend in all directions, just like our backbone.

✂ **(Optional) Human Body Project** – This week, have the students add the skeleton to the back of the body outline on pp. 105-106; pictures for this project are on p. 104. As you add the part, do a quick oral review of the skeletal system by asking, "Do you remember what the skeleton does?".

To-Do List

Essential

- ☐ Read the appropriate reading assignment.
- ☐ Define skeleton.
- ☐ Complete the lapbook or notebook assignments.
- ☐ Do the "Backbone Model" activity.

Optional

- ☐ Get one or more of the library books to read.
- ☐ Fill out a lab report sheet for one of the activities.
- ☐ Do the "Human Body Project" activity.

Lesson 4: Muscular System

Information

Reading Assignments

- **Younger Students** - *DK First Human Body Encyclopedia* pp. 22-23 Moving Joints, pp. 24-25 The Body's Muscles, pp. 26-27 How Muscle's Work
- **Older Students** - *Kingfisher Science Encyclopedia* p. 105 Joints, pp. 106-107 Muscles and Movement

(Optional) Books from the Library

- *Bend and Stretch: Learning About Your Bones and Muscles (Amazing Body)* by Pamela Hill Nettleton
- *The Mighty Muscular-Skeletal System: How Do My Bones and Muscles Work?* by John Burstein
- *Your Muscles (Your Body)* by Anne Ylvisaker

Notebooking

Vocabulary

Go over the following word with your students. Then, have them create a flashcard or copy the definition into the glossary.

- **Muscle** – A type of tissue that makes the bones of your body move and that is controlled by your brain. (Flashcard LT p. 21; Glossary NB p. 62)

Writing Instructions

- **Lapbook** – Have the students cut out the "Muscular System" tab-book on LT p.12 and color the cover. Then, have them tell you the purpose (or job) of the muscular system and record it on the purpose page. After that, have the students tell you what they have learned about muscles and how muscles work. Write it for them on the respective pages. Finally, staple the pages together and glue the tab-book into the lapbook.
- **Notebook** – Have the students dictate, copy, or write one to four sentences for the joints, muscles, and how muscles work on the muscular system notebooking page on NB p. 9.

Hands-on Science

Coordinating Activities

- **Model Joints** – Have the students make a model of both a pivot joint and a hinge joint. You will need a brad, card stock, scissors, and a hole punch to make a pivot joint. You will need two lengths of 1'x2' boards, a hinge, a screwdriver, and several screws to make a hinge joint. See the Appendix p. 102 for directions for this project. (**Note** – There are also directions for an optional ball and socket joint. If you choose to do this joint, you will also need one length of ½" plastic pipe, one length of ¾" plastic pipe, one ½" plastic pipe tee, one ¾" plastic pipe tee, a pipe saw, and two rubber bands to make a ball and socket joint.)

✂ **(Optional) Human Body Project** – This week, have the students add the arm muscle to the front of the body outline on pp. 105-106; pictures for this project are on p. 104. As you add the part, do a quick oral review of the muscular system by asking, "Do you remember what the muscles do for the body?".

To-Do List

Essential

- ☐ Read the appropriate reading assignment.
- ☐ Define muscle.
- ☐ Complete the lapbook or notebook assignments.
- ☐ Do the "Model Joints" activity.

Optional

- ☐ Get one or more of the library books to read.
- ☐ Fill out a lab report sheet for one of the activities.
- ☐ Do the "Human Body Project" activity.

Lesson 5: Nervous System

Information

Reading Assignments

- **Younger Students** - *DK First Human Body Encyclopedia* pp. 30-31 Headquarters, pp. 32-33 Network of Nerves
- **Older Students** - *Kingfisher Science Encyclopedia* pp. 108-109 The Brain and Nervous System

(Optional) Books from the Library

- *The Nervous System (New True Books: Health)* by Christine Taylor-Butler
- *The Brain: Our Nervous System* by Seymour Simon
- *Brain, Nerves, and Senses (Understanding the Human Body)* by Steve Parker
- *You've Got Nerve!: The Secrets of the Brain and Nerves (The Gross and Goofy Body)* by Melissa Stewart

Notebooking

Vocabulary

Go over the following word with your students. Then, have them create a flashcard or copy the definition into the glossary.

- **Neuron** – A nerve cell that makes up the nervous system and carries electrical messages throughout the body. (Flashcard LT p. 22; Glossary NB p. 62)

Writing Instructions

- **Lapbook** – Have the students cut out the "Nervous System" tab-book on LT p. 13 and color the cover. Then, have them tell you the purpose (or job) of the nervous system and record it on the purpose page. After that, have the students tell you what they have learned about the brain and spinal cord. Write it for them on the respective pages. Finally, staple the pages together and glue the tab-book into the lapbook.

- **Notebook** – Have the students dictate, copy, or write one to four sentences for the brain, nervous system, and sleep on the nervous system notebooking page on NB p. 10.

Hands-on Science

Coordinating Activities

- **Testing Reflexes** – Begin by sharing with the students about reflexes. You could say, "Reflexes are used to protect the body without us having to think about what is happening. Reflexes get us away from objects that might hurt us, before they hurt us. For example, if you put your hand on a hot stove, you immediately remove your hand BEFORE the message, 'Hey, my hand is on a hot, burning stove', gets to your brain." Then, suddenly slam a book on a table to create a loud noise. The students should have twitched, moved their heads, blinked their eyes, put their hands up, or screamed. Their response is their body's automatic reflex to loud noises. Another demonstration of these built-in capabilities is the blink reflex. Have the students stand behind a see-through barrier like a window or a wire screen. Throw a cotton ball at the student. Did they blink? Probably and their response is known as the blink reflex which serves to protect our eyes from damage.

✂ **(Optional) Human Body Project** – This week, have the students add the brain to the front of the body outline and the spinal cord to the back of the body outline on pp. 105-106; pictures for this project are on p. 104. As you add the part, do a quick oral review of the nervous system by asking, "Do you remember what the brain and spinal cord do?".

To-Do List

Essential

- ☐ Read the appropriate reading assignment.
- ☐ Define neuron.
- ☐ Complete the lapbook or notebook assignments.
- ☐ Do the "Testing Reflexes" activity.

Optional

- ☐ Get one or more of the library books to read.
- ☐ Fill out a lab report sheet for one of the activities.
- ☐ Do the "Human Body Project" activity.

Lesson 6: The Five Senses

Information

Reading Assignments

- **Younger Students** - *DK First Human Body Encyclopedia* pp. 38-39 How we see, pp. 44-45 Listen Up, pp. 36-37 Taste & Smell, pp. 34-35 Touchy Feely
- **Older Students** - *Kingfisher Science Encyclopedia* p. 112 Touch, p. 113 Taste and Smell, p. 114-115 Eyes and Seeing, pp. 116-117 Ears, Hearing and Balance

(Optional) Books from the Library

- *Look, Listen, Taste, Touch, and Smell: Learning About Your Five Senses (Amazing Body)* by Pamela Hill Nettleton
- *My Five Senses Big Book (Let's-Read-And-Find...* by Margaret Miller and Aliki
- *The Magic School Bus Explores the Senses* by Joanna Cole
- *The Listening Walk* by Paul Showers

Notebooking

Vocabulary

Go over the following word with your students. Then, have them create a flashcard or copy the definition into the glossary.

- **Senses** – The ability of the body to take in and respond to information from its surroundings. (Flashcard LT p. 22; Glossary NB p. 64)

Writing Instructions

- **Lapbook** – Have the students color each of the five senses pictures on the "Five Senses Accordion" fold-book on LT p. 4. Have the students tell you what they have learned about each of the senses. Then, write for them their favorite piece of information on the fold-book. Finally, fold the book like an accordion and glue the back into the lapbook.
- **Notebook** – Have the students dictate, copy, or write one to four sentences for each of the senses on the five senses notebooking page on NB pp. 12-13.

Hands-on Science

Coordinating Activities

- **Blind Food Test** – Choose several different types of food from your kitchen and place each of them on a separate paper plate. Blindfold the students and have them try to identify the item by using their remaining four senses—touch, smell, taste, and hearing. Write down their answer on the plate for each item. At the end, show the students the items and their answers so they can see how many they got right.

To-Do List

Essential

- ☐ Read the appropriate reading assignment.
- ☐ Define senses.
- ☐ Complete the lapbook or notebook assignments.
- ☐ Do the "Blind Food Test" activity.

Optional

- ☐ Get one or more of the library books to read.
- ☐ Fill out a lab report sheet for one of the activities.

Lesson 7: Circulatory System

Information

Reading Assignments

- **Younger Students** - *DK First Human Body Encyclopedia* pp. 50-51 Boom, Boom, pp. 48-49 Blood Flow, pp. 54-55 Blood Cells
- **Older Students** - *Kingfisher Science Encyclopedia* pp. 120-121 Heart and Circulation, p. 122 Blood

(Optional) Books from the Library

- *The Heart: Our Circulatory System* by Seymour Simon
- *The Circulatory Story* by Mary K. Corcoran
- *The Amazing Circulatory System: How Does My Heart Work?* by John Burstein
- *Hear Your Heart (Let's-Read-and-Find... Science 2)* by Paul Showers and Holly Keller
- *The Magic School Bus Has a Heart* by Anne Capeci and Carolyn Bracken

Notebooking

Vocabulary

Go over the following word with your students. Then, have them create a flashcard or copy the definition into the glossary.

- **Blood vessel** – A tube that carries blood through the body. (Flashcard LT p. 23; Glossary NB p. 57)

Writing Instructions

- **Lapbook** – Have the students cut out the "Circulatory System" tab-book on LT pp. 15-16 and color the cover. Then, have them tell you the purpose (or job) of the circulatory system and record it on the purpose page. After that, have the students tell you what they have learned about the heart and blood vessels. Write it for them on the respective pages. Next, have the students label the different cells on the blood page. Finally, staple the pages together and glue the tab-book into the lapbook.
- **Notebook** – Have the students dictate, copy, or write one to four sentences for the heart, blood vessels, and blood on the circulatory system notebooking page on NB p. 11.

Hands-on Science

Coordinating Activities

- **Heartbeat** – Have the students use a paper towel tube to listen to each other's heartbeats.

Begin by having them place the tube over their partner's heart and put their ear up to the other end of the tube. Then, have them count the number of beats in one minute. To take it a step further, have them compare the number of beats an adult's heart makes in one minute to how many a child's heart makes in one minute. (*A child's heart generally beats faster than an adult's one.*)

(Optional) Human Body Project – This week, have the students add the heart to the front of the body outline on pp. 105-106; pictures for this project are on p. 104. As you add the part, do a quick oral review of the circulatory system by asking, "Do you remember what the heart does?".

To-Do List

Essential

- ☐ Read the appropriate reading assignment.
- ☐ Define heartbeat.
- ☐ Complete the lapbook or notebook assignments.
- ☐ Do the "Heartbeat" activity.

Optional

- ☐ Get one or more of the library books to read.
- ☐ Fill out a lab report sheet for one of the activities.
- ☐ Do the "Human Body Project" activity.

Lesson 8: Respiratory System

Information

Reading Assignments

- **Younger Students** - *DK First Human Body Encyclopedia* pp. 62-63 Air & Oxygen, pp. 60-61 Air Bags, pp. 64-65 Making Sounds
- **Older Students** - *Kingfisher Science Encyclopedia* pp. 124-125 Lungs and Breathing

(Optional) Books from the Library

- *The Respiratory System: Why Do I Feel Out of Breath?* by Sue Barraclough
- *How Do Your Lungs Work? (Rookie Read-About Health)* by Don L. Curry
- *The Remarkable Respiratory System: How Do My Lungs Work?* by John Burstein
- *Breathe In, Breathe Out: Learning About Your Lungs (Amazing Body)* by Pamela Hill Nettleton

Notebooking

Vocabulary

Go over the following word with your students. Then, have them create a flashcard or copy the definition into the glossary.

- **Alveoli** – Tiny air bags found in your lungs. (Flashcard LT p. 23; Glossary NB p. 56)

Writing Instructions

- **Lapbook** – Have the students cut out the "Respiratory System" tab-book on LT p. 17 and color the cover. Then, have them tell you the purpose (or job) of the respiratory system and record it on the purpose page. After that, have the students tell you what they have learned about the lungs and alveoli. Write it for them on the respective pages. Finally, staple the pages together and glue the tab-book into the lapbook.
- **Notebook** – Have the students dictate, copy, or write one to four sentences for the lungs, alveoli, and breathing on the respiratory system notebooking page on NB p. 14.

Hands-on Science

Coordinating Activities

- **Inhalation versus Exhalation** – Have the students complete the Inhalation versus Exhalation worksheet on p. 103 of the Appendix. They should label each process and what happens during them. If you would like to see these processes in action, have the students make their own set of model lungs using the directions from the following website:
  - Model Lung Tutorial by MEL Science - https://youtu.be/fybV8zIGyu8

✂ **(Optional) Human Body Project** – This week, have the students add the lungs to the front of the body outline on pp. 105-106; pictures for this project are on p. 104. As you add the part, do a quick oral review of the respiratory system by asking, "Do you remember what the lungs does?".

To-Do List

Essential

- ☐ Read the appropriate reading assignment.
- ☐ Define alveoli.
- ☐ Complete the lapbook or notebook assignments.
- ☐ Do the "Heartbeat" activity.

Optional

- ☐ Get one or more of the library books to read.
- ☐ Fill out a lab report sheet for one of the activities.
- ☐ Do the "Human Body Project" activity.

Lesson 9: Digestive System

Information

Reading Assignments

- **Younger Students** - *DK First Human Body Encyclopedia* pp. 82-83 Digestive System, pp. 84-85 Chew it Over, pp. 88-89 Inside Intestines
- **Older Students** - *Kingfisher Science Encyclopedia* pp. 128-129 Digestion, p. 127 Teeth

(Optional) Books from the Library

- *What Happens to a Hamburger? (Let's-Read-and-Find... Science 2)* by Paul Showers
- *Where Does Your Food Go? (Rookie Read-About Health)* by Wiley Blevins
- *The Dynamic Digestive System: How Does My Stomach Work?* by John Burstein
- *Guts: Our Digestive System* by Seymour Simon

Notebooking

Vocabulary

Go over the following word with your students. Then, have them create a flashcard or copy the definition into the glossary.

- **Digestion** – The process by which your food is broken down. (Flashcard LT p. 24; Glossary NB p. 59)

Writing Instructions

- **Lapbook** – Have the students cut out the "Digestive and Urinary System Accordion" fold-book (multi-week) on LT p. 18 and color the pictures. Then, have them tell you the purpose (or job) of the digestive system and record it on the correct block. After that, have the students tell you what they have learned about the teeth and intestines. Write it for them on the respective blocks. The students will finish the fold-book next week.

- **Notebook** – Have the students dictate, copy, or write one to four sentences for the digestive system, teeth, and intestines on the digestive system notebooking page on NB p. 15.

Hands-on Science

Coordinating Activities

- **Stomach Model** – Have the students make a model of the stomach. You will need a large Ziploc bag, bread and coke. Have the students open the bag and add several pieces of bread. Have them mush up the bread with their hands, simulating what happens when they chew up the bread. Then, have them add 1 cup of coke to simulate stomach acid. Continue to squeeze and shake the bag around to see what happens to the bread. (*The bread should break down into a pulp, just like what happens to our food in the stomach.*)

✂ **(Optional) Human Body Project** – This week, have the students add the digestive system to the front of the body outline on pp. 105-106; pictures for this project are on p. 104. As you add the part, do a quick oral review of the digestive system by asking, "How does food move through the digestive system?".

To-Do List

Essential

- ☐ Read the appropriate reading assignment.
- ☐ Define digestion.
- ☐ Complete the lapbook or notebook assignments.
- ☐ Do the "Stomach Model" activity.

Optional

- ☐ Get one or more of the library books to read.
- ☐ Fill out a lab report sheet for one of the activities.
- ☐ Do the "Human Body Project" activity.

Lesson 10: Urinary System

Information

Reading Assignments

- **Younger Students** - *DK First Human Body Encyclopedia* pp. 90-91 Waterworks, pp. 92-93 Stretchy Bladder
- **Older Students** - *Kingfisher Science Encyclopedia* pp. 131 Waste Disposal

(Optional) Books from the Library

- *The Digestive and Excretory Systems (The Human Body Library)* by Susan Dudley Gold
- *Learning about the Digestive and Excretory Systems (Learning about the Human Body Systems)* by Susan Dudley Gold
- *My Messy Body (Body Works)* by Liza Fromer, Francine and Joe Weissmann

Notebooking

Vocabulary

Go over the following word with your students. Then, have them create a flashcard or copy the definition into the glossary.

- **Kidney** – The organ in the body responsible for removing waste from itself and regulating the body's fluid levels. (Flashcard LT p. 24; Glossary NB p. 61)

Writing Instructions

- **Lapbook** – Have the students cut out the "Digestive and Urinary System" accordion fold-book on LT p. 18 and color the pictures. Then, have them tell you the purpose (or job) of the urinary system and record it on the correct block. After that, have the students tell you what they have learned about the kidneys. Write it for them on the respective block. Finally, fold the book like an accordion and glue the back into the lapbook.
- **Notebook** – Have the students dictate, copy, or write one to four sentences for the urinary system and bladder on the urinary system notebooking page on NB p. 16.

Hands-on Science

Coordinating Activities

- **Stretchy Bladder** – You will need a balloon, funnel, and water for this activity. Have the students use the funnel to fill the balloon with water one cup at a time. Observe how the balloon stretches to accommodate the water, but at some point it reaches full capacity. Then have the students pour the water out and see if the balloon has changed. (*The students should* *see that the balloon can contain several cups of water, but when the water is poured out it returns to its original shape. This is just like the bladder in the human body. It stretches to store the urine produced by the body and returns to its original shape once it is emptied.)*

✂ **(Optional) Human Body Project** – This is the last week for this project. Have the students add the urinary system to the front of the body outline on pp. 105-106; pictures for this project are on p. 104. As you add the part, do a quick oral review of the urinary system by asking, "What are this part and what does it do?".

To-Do List

Essential

- ☐ Read the appropriate reading assignment.
- ☐ Define kidney.
- ☐ Complete the lapbook or notebook assignments.
- ☐ Do the "Stretchy Bladder" activity.

Optional

- ☐ Get one or more of the library books to read.
- ☐ Fill out a lab report sheet for one of the activities.
- ☐ Do the "Human Body Project" activity.

Lesson 11: Immune System

Information

Reading Assignments

- **Younger Students** - *DK First Human Body Encyclopedia* pp. 74-75 Germs, pp. 76-77 Body Defense
- **Older Students** - *Kingfisher Science Encyclopedia* p. 136 Bacteria and Viruses, p. 137 The Immune System, p. 138 Diseases

(Optional) Books from the Library

- *The Immune System Your Magic Doctor: A Guide to the Immune System for the Curious of All Ages* by Helen Garvy and Dan Bessie
- *Body Warriors: The Immune System* by Lisa Trumbauer
- *Our Immune System (Our Bodies (Discovery Library))* by Susan Thames
- *Germs* by Ross Collins

Notebooking

Vocabulary

Go over the following word with your students. Then, have them create a flashcard or copy the definition into the glossary.

- **Bacteria** – A group of microscopic organisms that can cause diseases. (Flashcard LT p. 25; Glossary NB p. 57)

Writing Instructions

- **Lapbook** – Have the students cut out the "Immune System" tab-book on LT p. 19 and color the cover. Then, have them tell you the purpose (or job) of the immune system and record it on the purpose page. After that, have the students tell you what they have learned about germs and the body's defenses. Write it for them on the respective pages. Finally, staple the pages together and glue the tab-book into the lapbook.
- **Notebook** – Have the students dictate, copy, or write one to four sentences for the germs, body defenses, and allergies on the immune system notebooking page on NB p. 17.

Hands-on Science

Coordinating Activities

- **Germ Test** – Have the students determine which room has the most germs. You will need sterile cotton swabs, petri dishes, and agar. Use the cotton swabs to test different areas in several rooms in the house. See the following website for directions for this process:
  - How to Test for Bacteria on Kitchen Counters - http://www.ehow.com/how_6381785_test-bacteria-kitchen-counters.html

Review Sheet

Human Body Review Sheet Answers

1. Billions
2. True
3. Dermis, Epidermis
4. False (*Hair is made up of dead cells.*)
5. Supports your body, Allows you to move, Protects certain organs
6. Brain
7. Brain
8. True
9. Neuron
10. False (*The brain is the command center for the nervous system.*)
11. Nose-smell, Ear-hearing, Eye-sight, Skin-feel, Tongue-taste
12. Pumps
13. Blood vessels
14. True
15. Oxygen, Carbon dioxide
16. Broken down
17. Your teeth
18. Waste
19. True
20. Microscopic
21. Immune

To-Do List

Essential

- ☐ Read the appropriate reading assignment.
- ☐ Define bacteria.
- ☐ Complete the lapbook or notebook assignments.
- ☐ Do the "Germ Test" activity.

Optional

- ☐ Get one or more of the library books to read.
- ☐ Fill out a lab report sheet for one of the activities.
- ☐ Complete the Human Body Review Sheet. (pp. 122-125)

SCIENCE CHUNKS

Weather Unit

Lesson 1: The Sun and Atmosphere

Information

Reading Assignments

- **Younger Students** - *DK Children's Encyclopedia* p. 33 Atmosphere, p. 247 Sun
- **Older Students** - *Kingfisher Science Encyclopedia* pp. 10-11 Earth's Atmosphere, pp. 394-395 The Sun

(Optional) Books from the Library

- *The Sun: Our Nearest Star (Let's-Read-and-Find Out)* by Franklyn M. Branley and Edward Miller
- *The Sun* by Seymour Simon
- *Grandpa, What is Air? (Popular Science for Children Ages 4-8, Books' Series)* by Daniel Levy
- *Day and Night (First Step Nonfiction: Discovering Nature's Cycles)* by Robin Nelson
- *Day and Night (Patterns in Nature)* by Margaret Hall and Jo Miller
- *What Makes Day and Night (Let's-Read-and-Find Out Science 2)* by Franklyn M. Branley and Arthur Dorros

Notebooking

Vocabulary

Go over the following word with your students. Then, have them create a flashcard or copy the definition into the glossary.

- **Atmosphere** – A layer of gas that surrounds a planet. (Flashcard LT p. 36; Glossary NB p. 56)

Writing Instructions

- **Lapbook** – Have the students begin the Weather lapbook by cutting out and coloring the cover on LT p. 28.
- **Lapbook** – Have the students complete the Sun mini-book on LT p. 29. Have them cut out the pages and color the cover. Next, have the students tell you what they have learned about how the sun affects Earth's weather. Then, have them fold the book and glue it into their lapbook.
- **Lapbook** – Have the students complete the Layers of the Atmosphere card on LT p. 30. Have the younger students cut out and glue the completed card into their lapbook.
- **Lapbook** – Have the students cut out and glue the vocabulary pocket on LT p. 36 into their lapbook.
- **Notebook** – Have the students dictate, copy, or write one to four sentences about the sun and Earth's atmosphere on the sun and atmosphere notebooking page on NB p. 20.

Hands-on Science

Coordinating Activities

✂ **Rain Gauge** – This project will be completed over the remainder of this unit. This week, you will need a plastic water bottle, duct tape, a permanent marking pen, small marbles or rocks, and a ruler. Begin by cutting the top off of the plastic bottle at the point where the diameter of the bottle starts to decrease. Keep the top portion of the bottle that was cut off and cover all the cut edges with duct tape to hide the sharp edges. On the bottom section of the bottle use a ruler to make a scale of horizontal lines in centimeters with the permanent marker. Have the students make the first line four centimeters from the bottom. Then, have them add a line layer of marbles or rocks in the bottom portion of the bottle to steady the base. Next, have them pour water into the bottle up to the first line of their scale. Finally, have them turn the top portion of the bottle upside down and place it inside the opening of the bottom portion to form a funnel-shaped lid. Now the students are ready to begin measuring the rain with their bottle. The rain gauge chart on pp. 107-108 of the Appendix has both a predicted rain amount and an actual rain amount. After reviewing the weather report for the week, have the students predict how much they think it is going to rain each day. You can choose to have them record the rain for 5 or 7 days a week. Either way, they will be recording their observations in a similar fashion over the remaining weeks in this unit.

✂ **(Optional) Atmosphere Collage** – Have the student make an atmosphere collage that depicts the different object one can find in the various layers of the atmosphere. Use the reference books for ideas of what to draw. Have the older students write a sentence about what each of the layers are like.

To-Do List

Essential

- ☐ Read the appropriate reading assignment.
- ☐ Define atmosphere.
- ☐ Complete the lapbook or notebook assignments.
- ☐ Do the "Rain Gauge" activity.

Optional

- ☐ Get one or more of the library books to read.
- ☐ Fill out a lab report sheet (p. 101) for one of the activities.
- ☐ Do the "Atmosphere Collage" activity

Lesson 2: Seasons and Climate

Information

Reading Assignments

- **Younger Students** - *DK Children's Encyclopedia* p. 221 Seasons
- **Older Students** - *Kingfisher Science Encyclopedia* pp. 36-37 Climate
  Note - If your older students don't know the four seasons, please share them with the students this week.

(Optional) Books from the Library

- *Sunshine Makes the Seasons (Let's-Read-and-Find Out Science 2)* by Franklyn M. Branley and Michael Rex
- *Watching the Seasons (Welcome Books)* by Edana Eckart
- *The Reasons for Seasons* by Gail Gibbons
- *Tree For All Seasons (Avenues)* by Robin Bernard
- *On the Same Day in March: A Tour of the World's Weather* by Marilyn Singer and Frane Lessac
- *Weather and Climate: Geography Facts and Experiments (Young Discoverers Series)* by Barbara Taylor

Notebooking

Vocabulary

Go over the following word with your students. Then, have them create a flashcard or copy the definition into the glossary.

- **Climate** – The average weather in an area over a given period of time. (Flashcard LT p. 37; Glossary NB p. 58)

Writing Instructions

- **Lapbook** – Have the students complete the Four Seasons Wheel on LT p. 31. Have them cut out the two wheels and color the pictures. Then, have them tell you the typical weather for your area in each of the seasons. Write this information down on their wheels. Once the students are done, assemble the wheel-book with a brad fastener and glue it into the lapbook.
- **Notebook** – Have the students dictate, copy, or write one to four sentences about climate on the climate notebooking page on NB p. 21.

Hands-on Science

Coordinating Activities

- **Seasons Tree** – Have the students make a tree from a season of their choice. You will need a lump of air dry clay, several brown pipe cleaners, and felt sheets of either green, red, orange, or yellow. Have the students begin by shaping the pipe cleaners for the trunk and branches. Once they are finished, have them secure the tree into a lump of clay as a base. Then, have them pick a season and cut out felt leaves in the right color. In other words, if they chose to make a spring tree, they would use light green felt for their leaves. Have the students glue these onto the tree. Finally, have them decorate their tree with things like flowers, bird's nests, and fruit or nuts. For example, for the spring tree the students could add small pieces of crushed tissue paper for flowers.

- **(Optional) Seasons Book** – Have the students make a book about the seasons. See the following blog post for directions and a free printable:
  - https://elementalscience.com/blogs/science-activities/seasons-hawk-talons
- **(Optional) Rain Gauge** – Have the students continue to check and measure the rainfall using the rain gauge they made in lesson 1.

To-Do List

Essential

- ☐ Read the appropriate reading assignment.
- ☐ Define climate.
- ☐ Complete the lapbook or notebook assignments.
- ☐ Do the "Seasons Tree" activity.

Optional

- ☐ Get one or more of the library books to read.
- ☐ Fill out a lab report sheet for one of the activities.
- ☐ Do the "Seasons Book" or "Rain Gauge" activity.

Lesson 3: Weather and Wind

Information

Reading Assignments

- **Younger Students** - *DK Children's Encyclopedia* p. 271 Weather
  Note - If you choose to use this resource, you will need to learn about tornadoes by reading a book from the library or by watching the following video:
  - https://www.youtube.com/watch?v=aacHWoB7cmY
- **Older Students** - *Kingfisher Science Encyclopedia* pp. 44-45 Wind, Storms, and Floods

(Optional) Books from the Library

- *Oh Say Can You Say What's the Weather Today?* by Tish Rabe and Aristides Ruiz
- *Weather* by Seymour Simon
- *Weather Words and What They Mean* by Gail Gibbons
- *Wind* by Marion Dane Bauer and John Wallace
- *Feel the Wind (Let's-Read-and-Find Out Science 2)* by Arthur Dorros
- *The Wind Blew* by Pat Hutchins
- *Tornadoes (Blastoff Readers. Level 4)* by Anne Wendorff
- *Tornadoes!* by Gail Gibbons
- *Tornado Alert* by Franklyn M. Branley and Giulio Maestro

Notebooking

Vocabulary

Go over the following word with your students. Then, have them create a flashcard or copy the definition into the glossary.

- **Tornado** – A spinning funnel of wind. (Flashcard LT p. 37; Glossary NB p. 67)

Writing Instructions

- **Lapbook** – Have the students begin the Weather tab-book on LT p. 32. Have them cut out the cover and Wind-tab pages. Next, have them color the cover and tell you what they have learned about wind and tornadoes. Write what they say on the Wind-tab page. Then, set aside the pages as you will continue to work on this mini-book for the rest of the unit.
- **Notebook** – Have the students dictate, copy, or write one to four sentences about weather, storms, and floods on the weather and wind notebooking page on NB p. 22.

Hands-on Science

Coordinating Activity

- ✂ **Wind Race** – Have a wind race with the students. You will need a paper or cotton ball and a straw for each of the students. Have them sit at one end of the table with their ball directly in front of them. When you say go, have the students use their straws to blow their ball to the other end of the table. The first person to do so is the winner.
- ✂ **(Optional) Rain Gauge** – Have the students continue to check and measure the rainfall using the rain gauge they made in lesson 1.

To-Do List

Essential

- ☐ Read the appropriate reading assignment.
- ☐ Define tornado.
- ☐ Complete the lapbook or notebook assignments.
- ☐ Do the "Wind Race" activity.

Optional

- ☐ Get one or more of the library books to read.
- ☐ Fill out a lab report sheet for one of the activities.
- ☐ Do the "Rain Gauge" activity.

Lesson 4: Water Cycle and Clouds

Information

Reading Assignments

- **Younger Students** - *DK Children's Encyclopedia* p. 64 Clouds, p. 270 Water Cycle
- **Older Students** - *Kingfisher Science Encyclopedia* pp. 40-41 Clouds and Fog
  Note - You can also have your older students reread the section on the water cycle at the top of p. 37 in the *Kingfisher Science Encyclopedia.*

(Optional) Books from the Library

- *The Water Cycle* by Helen Frost
- *The Water Cycle (Earth and Space Science)* by Craig Hammersmith
- *The Water Cycle (Water All Around)* by Rebecca Olien and Ted Williams
- *Clouds (Let's-Read-and-Find Out Science 1)* by Anne Rockwell and Frane Lessac
- *The Cloud Book* by Tomie dePaola
- *Clouds (Ready-to-Reads)* by Marion Dane Bauer and John Wallace
- *Fog and Mist (Watching the Weather)* by Elizabeth Miles
- *Fog (Weather)* by Helen Frost

Notebooking

Vocabulary

Go over the following word with your students. Then, have them create a flashcard or copy the definition into the glossary.

- **Vapor** – Tiny droplets of water in the air. (Flashcard LT p. 38; Glossary NB p. 67)
- **Natural cycle** – A naturally occurring event that happens over and over again in the same order. (Flashcard LT p. 38; Glossary NB p. 62)

Writing Instructions

- **Lapbook** – Have the students add the Water Cycle sheet on LT p. 35. Have the younger students color, cut out, and glue the completed sheet into their lapbook. Have the older students cut out the blank sheet, label the cycle, color it, and glue it into their lapbook.

- **Lapbook** – Have the students continue to work on the Weather tab-book on LT p. 33. This week, have them cut out the Cloud-tab page and color the image. Next, have the students tell you what they have learned about the clouds and the shapes of clouds. Write what they say on the Cloud-tab page. Then, set aside the pages as you will continue to work on this mini-book for the rest of the unit.

- **Notebook** – Have the students dictate, copy, or write about clouds and the steps of the water cycle on the water cycle and clouds notebooking page on NB p. 22. The answers for this can be found on the next page.

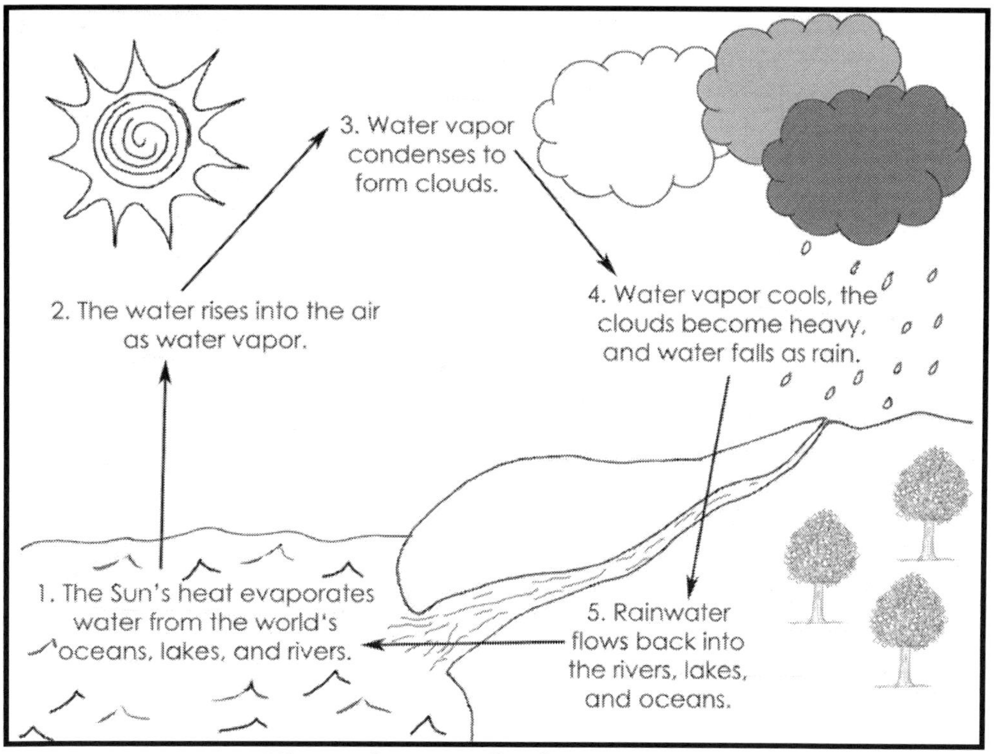

Hands-on Science

Coordinating Activity

- **Fog** – Have the students see how fog forms. You will need a glass jar or cup, plastic wrap, rubber band, and ice cubes. Begin by filling the glass jar with hot water and let it sit for several minutes. Now, pour out almost all of the hot water. (**Note**—Leave about 1 to 2 inches of water in the bottom.) Quickly cover the jar loosely with plastic wrap and use a rubber band to secure it. (**Note**—There needs to be a bit of a pocket formed by the plastic wrap at the top of the jar in which to place the ice cubes.) Place the ice cubes on the plastic wrap and observe what happens. The students should see that fog develops in the jar just above the water. This is because when the cold air from the ice meets with the warm air in the jar, the water vapor inside condenses. This condensation is known as fog.
- **(Optional) Rain Gauge** – Have the students continue to check and measure the rainfall using the rain gauge they made in lesson 1.

To-Do List

Essential

- ☐ Read the appropriate reading assignment.
- ☐ Define vapor and natural cycle.
- ☐ Complete the lapbook or notebook assignments.
- ☐ Do the "Fog" activity.

Optional

- ☐ Get one or more of the library books to read.
- ☐ Fill out a lab report sheet for one of the activities.
- ☐ Do the "Rain Gauge" activity.

Lesson 5: Storms and Rain

Information

Reading Assignments

- **Younger Students** - *DK Children's Encyclopedia* p. 246 Storms
- **Older Students** - *Kingfisher Science Encyclopedia* pp. 38-39 Rain and Snow

(Optional) Books from the Library

- *Rain!* by Linda Ashman and Christian Robinson
- *Down Comes the Rain (Let's-Read-and-Find Out Science 2)* by Franklyn M. Branley and James Graham Hale
- *Rain (Ready-to-Reads)* by Marion Dane Bauer and John Wallace
- *National Geographic Readers: Storms!* by Miriam Goin
- *Hurricanes!* by Gail Gibbons
- *The Magic School Bus Inside A Hurricane* by Joanna Cole and Bruce Degen
- *Hurricanes (Blastoff Readers Level 4)* by Kay Manolis
- *Weather Forecasting* by Gail Gibbons
- *What Will the Weather Be? (Let's-Read-and-Find Out Science 2)* by Lynda DeWitt and Carolyn Croll

Notebooking

Vocabulary

Go over the following word with your students. Then, have them create a flashcard or copy the definition into the glossary.

- **Precipitation** – Rain, snow, sleet, or hail that falls to the ground. (Flashcard LT p. 39; Glossary NB p. 64)
- **Thunderstorm** – A storm with thunder and lightning. (Flashcard LT p. 39; Glossary NB p. 66)

Writing Instructions

- **Lapbook** – Have the students continue to work on the Weather tab-book on LT p. 33. Have them cut out the Rain-tab page and color the image. Next, have the students tell you what they have learned about rain, storms, and hurricanes. Write what they say on the Rain-tab page. Then, set aside the pages as you will continue to work on this mini-book for the rest of the unit.

- **Notebook** – Have the students dictate, copy, or write one to four sentences about precipitation and thunderstorm on the storms and rain notebooking page on NB p. 24.

Hands-on Science

Coordinating Activity

- ✂ **Rain in a Glass** – Have the students create a rainstorm in a glass. You will need a clear glass, shaving cream, blue food coloring, and warm water. Fill the glass halfway with warm water and squirt a shaving cream cloud on top of the water. You can pack the shaving cream in a bit before adding an indention in the center so that the food coloring won't run over the "cloud". Now, slowly add 20 to 30 drops of blue food coloring in the center of the shaving cream cloud. Watch and wait for the food coloring "rain" to fall out of the bottom of the shaving cream "cloud".
- ✂ **(Optional) Rain Gauge** – Have the students continue to check and measure the rainfall using the rain gauge they made in lesson 1.

To-Do List

Essential

- ☐ Read the appropriate reading assignment.
- ☐ Define precipitation and thunderstorm.
- ☐ Complete the lapbook or notebook assignments.
- ☐ Do the "Rain in a Glass" activity.

Optional

- ☐ Get one or more of the library books to read.
- ☐ Fill out a lab report sheet for one of the activities.
- ☐ Do the "Rain Gauge" activity.

Lesson 6: Weather Forecasting

Information

Reading Assignments

- **Younger Students** - *DK Children's Encyclopedia* (No pages scheduled.)
  Note - If you choose to use this resource, you will need to learn about weather forecasting by reading a book from the library or by watching the following video:
  - https://www.youtube.com/watch?v=dLQ0lHpyZd8
- **Older Students** - *Kingfisher Science Encyclopedia* pp. 42-43 Weather Forecasting

(Optional) Books from the Library

- *A Kid's Guide to Weather Forecasting* by Baby Professor
- *Meteorology: Cool Women Who Weather Storms (Girls in Science)* by Karen Bush Gibson and Lena Chandhok

Notebooking

Vocabulary

There are no vocabulary words for this lesson.

Writing Instructions

- **Lapbook** – Have the students finish the Weather tab-book on LT p. 34. Have them cut out the Forecasting-tab page and color the image. Next, have the students tell you what they have learned about forecasting the weather. Write what they say on the Forecasting-tab page. Then, staple all the pages from the Weather tab-book together and glue the booklet into their lapbook.
- **Notebook** – Have the students dictate, copy, or write one to four sentences about forecasting the weather on the weather forecasting notebooking page on NB p. 25.

Hands-on Science

Coordinating Activity

- **Fake Snow** – Have the students make fake snow using shaving cream and cornstarch. Mix equal parts of the two ingredients together until desired consistency is reached.
- **(Optional) Rain Gauge** – Have the students continue to check and measure the rainfall using the rain gauge they made in lesson 1.

Review Sheet

Weather Review Sheet Answers

1. True
2. Sun's
3. Fall, Winter, Spring, Summer
4. True
5. All three factors should be circled.
6. False (*Changes in the weather occur in the atmosphere or air.*)
7. True
8. False (*Rain is a form of precipitation.*)
9. 1-C, 2-D, 3-B, 4-A, 5-E
10. Flood — An overflow of a large amount of water; Drought — A long period of dry conditions and little water
11. True

To-Do List

Essential

- ☐ Read the appropriate reading assignment.
- ☐ Complete the lapbook or notebook assignments.
- ☐ Do the "Fake Snow" activity.

Optional

- ☐ Get one or more of the library books to read.
- ☐ Fill out a lab report sheet for one of the activities.
- ☐ Do the "Rain Gauge" activity.
- ☐ Complete the Weather Review Sheet. (pp. 126-127)

SCIENCE CHUNKS

Stars Unit

Lesson 1: Stars

Information

Reading Assignments

- 🕮 **All Students** – *Glow in the Dark Constellations* pp. 6-8 An Introduction to Stargazing
  The reading assignments for this Science Chunks unit are a bit different from other ones as all the students will be reading from the same book. If you would like to add more information, we have included optional reading assignments from a public-domain book and from two of the encyclopedias used in other Science Chunks units.
- 🕮 **(Optional) Younger Students** – *DK Children's Encyclopedia* p. 242 Stars, p. 73 Constellations
- 🕮 **(Optional) Older Students** – *Usborne Science Encyclopedia* pp. 158-161 Stars

(Optional) Books from the Library

- 🕮 *The Sky Is Full of Stars (Let's-Read-and-Find... Science 2)* by Franklyn M. Branley and Felicia Bond
- 🕮 *Jump Into Science: Stars* by Steve Tomecek
- 🕮 *Stars! Stars! Stars!* by Bob Barner

Notebooking

Vocabulary

Go over the following words with your students. Then, have them create flashcards or copy the definitions into the glossary.

- **Constellation** – A group of stars that when viewed from Earth form the outline of an object or figure. (Flashcard LT p. 50; Glossary NB p. 58)
- **Star** – A huge ball of exploding gas. (Flashcard LT p. 50; Glossary NB p. 65)

Writing Instructions

- **Lapbook** – Have the students complete the Stars mini-book on LT p. 43. Have them cut out the mini-book and color the cover. Next, have the students tell you what they have learned about stars and stargazing and write this down for them in the mini-book. Then, have them fold the booklet, glue on the cover, and place it into the lapbook.
- **Lapbook** – Have the students cut out and glue the vocabulary pocket on LT p. 49 into their lapbook.
- **Notebook** – Have the students dictate, copy, or write one to four sentences about stars and stargazing on the stars notebooking page on NB p. 28.

Hands-on Science

Coordinating Activities

- **Shining Stars** – You will need black construction paper, a toothpick, tape, a flashlight, a large, clear bowl, and water. Have the students begin by using the toothpick to poke holes in the paper. (*These are the stars in the night sky.*) As they create the stars in their night sky, fill the bowl about three-quarters full of water. (*This will serve as the atmosphere.*) When the students are done, tape their night sky to the back of the bowl and head into a room without windows. Set the bowl on a flat surface and wait for the water to settle. Turn on the flashlight before turning off the lights in the room. Shine the flashlight on the back of the paper so that the light shines through the star holes and into the atmosphere bowl. Have the students observe the "stars" that are created. Gently tap the bowl so that the water begins to move, and have the students observe the changes to the "stars." (*The students should see that the stars appear to twinkle when the water moves because the light is being refracted, or bent, by the water. The same thing happens when light rays from the stars enter Earth's atmosphere, and this is the reason why the stars appear to twinkle at night.*)
- **(Optional) Life Cycle of a Star** – Have the students make a Life Cycle of a Star mobile. You will need a paper plate, string, two cotton balls, a small yellow pom-pom, a large red pom-pom, paint, and a white bead or sequin. Have them cut a paper plate into a spiral, punch a hole in the center of the spiral, and tie a string through it, so that they can hang up their mobile. Have the students use a pulled-out white cotton ball to make a stellar nebula with its cloud of dust and gas. Then, have them use a small yellow pom-pom for the average star and a large red pom-pom for the red giant. Next, have them paint a cotton ball purple, orange, and a bit of blue. Let it dry and pull it out of shape and use it for the planetary nebula. Finally, have the students use a small white bead or sequin the white dwarf. See the following post for example of a similar project:
  - https://elementalscience.com/blogs/science-activities/119870275-the-life-cycle-of-a-star-poster

To-Do List

Essential

- ☐ Read the appropriate reading assignment.
- ☐ Define constellation and star.
- ☐ Complete the lapbook or notebook assignments.
- ☐ Do the "Shining Stars" activity.

Optional

- ☐ Get one or more of the library books to read.
- ☐ Fill out a lab report sheet (p. 101) for one of the activities.
- ☐ Do the "Life Cycle of a Star" activity.

Lesson 2: Ursa Major and Ursa Minor

Information

Reading Assignments

- **All Students –** *Glow in the Dark Constellations* p. 10 Ursa Major, p. 12 Ursa Minor
- **(Optional) All Students** – *The Stars and Their Stories* pp. 7-15 The Two Bears
  The optional reading assignments for the remaining lessons all come from *The Stars and Their Stories* by Alice Griffith. This book is in the public domain and can be downloaded for free from Archive.org at the following link:
  - https://archive.org/details/starsandtheirst00unkngoog

(Optional) Books from the Library

- *The Big Dipper (Let's-Read-and-Find... Science 1)* by Franklyn M. Branley
- *The Heavenly Zoo: Legends and Tales of the Stars* by Alison Lurie (**Note—***This book is out of print, but it is wonderful if you can get your hands on it.*)

Notebooking

Writing Instructions

- **Lapbook** – Have the students complete the Two Bears flap-book on LT p. 44. Have them cut and color the cover. Next, have the students tell you what they have learned about Ursa Major and Ursa Minor, writing this for them on the respective tabs. Fold the tabs in and glue the cover sheet on the outside before gluing the mini-book into the lapbook.
- **Notebook –** Have the students dictate, copy, or write one to three sentences on the Ursa Major and Ursa Minor notebooking page on NB p. 29.

Star Labeling

For the remainder of the lessons, there are no vocabulary words. Instead, you can have the students label the key star(s) for the constellations they are studying in their lapbook or notebook. These stars will be shown in *Glow in the Dark Constellations.* Here is the star for this week's constellations:

- ★ Polaris (*Ursa Minor*)

Hands-on Science

Coordinating Activities

- **Planetarium** – Have the students make a homemade planetarium to view the constellations. (**Note –** *They will use this planetarium for the next four lessons.*) You will need foil, a toilet paper tube, a pin, a small flashlight, a rubber band, a Sharpie, and the constellation pictures for Ursa Major and Ursa Minor on p. 110. Begin by cutting the foil into two squares so that each square will fit around the end of the toilet paper tube and have enough overlap to be secured by the rubber band. Have the students use a pin and the constellation picture from the appendix to create two foil constellations, one for Ursa Major and one for Ursa Minor. Have them use a Sharpie to write the name of the constellation in the bottom corner of each piece of foil. Once they have the two constellations, have the students view each one by placing the foil template at one end of the toilet paper tube so that the constellation is centered in the middle of the tube. Then, secure the foil in place using the rubber band. Next, set the tube on a desk, place the flashlight in the other end of the tube, and aim the foil end of the tube toward a wall. Finally, turn off the lights, and turn on the flashlight to view the constellations. (**Note** – *If you want to take this a step further, have the students make a book of the constellations they looked at. They can make a constellation resist using the directions from the Starry Resist project from lesson 3 for the cover of the book. Then, have them make pages for Ursa Major and Ursa Minor. A template for this can be found in the appendix on p. 110.*)

- **(Optional) Pipe Cleaner Constellations** – Have the students make a pipe cleaner model of both Ursa Major and Ursa Minor. You will need pipe cleaners and star beads for this project. Have them use the constellation cards found in the appendix on p. 110 as a guide to create a pipe cleaner models. You can see how this project works here:
    - https://www.123homeschool4me.com/pipe-cleaner-constellations_61/

To-Do List

Essential

- ☐ Read the appropriate reading assignment.
- ☐ Complete the lapbook or notebook assignments.
- ☐ Label the star.
- ☐ Do the "Planetarium" activity.

Optional

- ☐ Get one or more of the library books to read.
- ☐ Fill out a lab report sheet for one of the activities.
- ☐ Do the "Pipe Cleaner Constellations" activity.

Lesson 3: Winter Constellations

Information

Reading Assignments

- **All Students** – *Glow in the Dark Constellations* p. 14 Orion, p. 16 Taurus
- **(Optional) All Students** – *The Stars and Their Stories* pp. 84-105 The Orion Group

(Optional) Books from the Library

- *The Heavenly Zoo: Legends and Tales of the Stars* by Alison Lurie

Notebooking

Writing Instructions

- **Lapbook** – Have the students complete the Winter Constellations tab-book on LT pp. 45. Have them cut and color the cover and the pages for Orion and Taurus. Next, have the students tell you what they have learned about the two constellations and write this for them on the respective pages. Then, staple the pages together and glue the mini-book into the lapbook.
- **Notebook** – Have the students dictate, copy, or write one to four sentences about Orion and Taurus on the Winter Constellations notebooking page on NB p. 30.

Star Labeling

Have the students label the following key stars on the constellations in their lapbook or notebook:

- Betelgeuse and Rigel (*Orion*)
- Aldebaran and the Pleiades (*Taurus*)

Hands-on Science

Coordinating Activities

- **Planetarium** – Have the students make a homemade planetarium to view the constellations. You will need the planetarium from last week, foil, a pin, a Sharpie, and the constellation pictures for Orion and Taurus on p. 110. Begin by cutting the foil into two squares so that each square will fit around the end of the toilet paper tube and have enough overlap to be secured by the rubber band. Have the students use a pin and the constellation picture from the appendix to create two foil constellations, one for Orion and one for Taurus. Have them use a Sharpie to write the name of the constellation in the bottom corner of each piece of foil. Once they have the two constellations, have the students view each one by placing the foil template at one end of the toilet paper tube so that the constellation is centered in the middle of the tube. Then, secure the foil in place using the rubber band. Next, set the tube on a desk, place the flashlight in the other end of the tube, and aim the foil end of the tube toward a wall. Finally, turn off the lights, and turn on the flashlight to view the constellations. (**Note** – *If you want to take this a step further, have the students make constellation pages for Orion and Taurus.*)

✂ **(Optional) Starry Resist** – Have the students make a night sky full of stars using resist painting. You will need gold star stickers, paper, and dark blue paint. Have the students place the gold stars in random places all over the paper. After they have done this, have them paint all over it with the dark blue paint. Once the paint is dry, have the students remove the star stickers to reveal their night sky of stars. (*If you want, you can have them recreate the Orion or the Taurus constellations.*)

To-Do List

Essential

- ☐ Read the appropriate reading assignment.
- ☐ Complete the lapbook or notebook assignments.
- ☐ Label the stars.
- ☐ Do the "Planetarium" activity.

Optional

- ☐ Get one or more of the library books to read.
- ☐ Fill out a lab report sheet for one of the activities.
- ☐ Do the "Starry Resist" activity.

Lesson 4: Spring Constellations

Information

Reading Assignments

- **All Students** – *Glow in the Dark Constellations* p. 18 Leo, p. 20 Virgo
- **(Optional) All Students** – *The Stars and Their Stories* pp. 139-145 The Two Lions and The Crab

(Optional) Books from the Library

- *The Heavenly Zoo: Legends and Tales of the Stars* by Alison Lurie

Notebooking

Writing Instructions

- **Lapbook** – Have the students complete the Spring Constellations tab-book on LT pp. 46. Have them cut and color the cover and the pages for Leo and Virgo. Next, have the students tell you what they have learned about the two constellations and write this for them on the respective pages. Then, staple the pages together and glue the mini-book into the lapbook.
- **Notebook** – Have the students dictate, copy, or write one to four sentences about Leo and Virgo on the Spring Constellations notebooking page on NB p. 31.

Star Labeling

Have the students label the following key stars on the constellations in their lapbook or notebook:

- Denebola and Regulus (*Leo*)
- Spica (*Virgo*)

Hands-on Science

Coordinating Activities

- **Planetarium** – Have the students make a homemade planetarium to view the constellations. You will need the planetarium from last week, foil, a pin, a Sharpie, and the constellation pictures for Leo and Virgo on p. 111. Begin by cutting the foil into two squares so that each square will fit around the end of the toilet paper tube and have enough overlap to be secured by the rubber band. Have the students use a pin and the constellation picture from the appendix to create two foil constellations, one for Leo and one for Virgo. Have them use a Sharpie to write the name of the constellation in the bottom corner of each piece of foil. Once they have the two constellations, have the students view each one by placing the foil template at one end of the toilet paper tube so that the constellation is centered in the middle of the tube. Then, secure the foil in place using the rubber band. Next, set the tube on a desk, place the flashlight in the other end of the tube, and aim the foil end of the tube toward a wall. Finally, turn off the lights, and turn on the flashlight to view the constellations. (**Note** – *If you want to take this a step further, have the students make constellation pages for Leo and Virgo.*)

- **(Optional) Edible Constellations** – Have the students make an edible model of the Leo and Virgo constellations they have studied. You will need marshmallows and pretzel sticks for this project. Have them use the constellation cards found in the appendix on p. 111 as a guide to create a stick model of the figures.

To-Do List

Essential

- ☐ Read the appropriate reading assignment.
- ☐ Complete the lapbook or notebook assignments.
- ☐ Label the stars.
- ☐ Do the "Planetarium" activity.

Optional

- ☐ Get one or more of the library books to read.
- ☐ Fill out a lab report sheet for one of the activities.
- ☐ Do the "Edible Constellations" activity.

Lesson 5: Summer Constellations

Information

Reading Assignments

- **All Students –** *Glow in the Dark Constellations* p. 22 Cygnus, p. 24 Scorpius
- **(Optional) All Students –** *The Stars and Their Stories* pp. 41-49 The Summer Triangle

(Optional) Books from the Library

- *The Heavenly Zoo: Legends and Tales of the Stars* by Alison Lurie

Notebooking

Writing Instructions

- **Lapbook –** Have the students complete the Summer Constellations tab-book on LT pp. 47. Have them cut and color the cover and the pages for Cygnus and Scorpius. Next, have the students tell you what they have learned about the two constellations and write this for them on the respective pages. Then, staple the pages together and glue the mini-book into the lapbook.
- **Notebook –** Have the students dictate, copy, or write one to four sentences about Cygnus and Scorpius on the Summer Constellations notebooking page on NB p. 32.

Star Labeling

Have the students label the following key stars on the constellations in their lapbook or notebook:

- Deneb and Sadr (*Cygnus*)
- Antares (*Scorpius*)

Hands-on Science

Coordinating Activities

- **Planetarium –** Have the students make a homemade planetarium to view the constellations. You will need the planetarium from last week, foil, a pin, a Sharpie, and the constellation pictures for Cygnus and Scorpius on p. 111. Begin by cutting the foil into two squares so that each square will fit around the end of the toilet paper tube and have enough overlap to be secured by the rubber band. Have the students use a pin and the constellation picture from the appendix to create two foil constellations, one for Cygnus and one for Scorpius. Have them use a Sharpie to write the name of the constellation in the bottom corner of each piece of foil. Once they have the two constellations, have the students view each one by placing the foil template at one end of the toilet paper tube so that the constellation is centered in the middle of the tube. Then, secure the foil in place using the rubber band. Next, set the tube on a desk, place the flashlight in the other end of the tube, and aim the foil end of the tube toward a wall. Finally, turn off the lights, and turn on the flashlight to view the constellations. (**Note** – *If you want to take this a step further, have the students make constellation pages for Cygnus and Scorpius.*)

✂ **(Optional) Geoboard Constellations** – Have the students make a geoboard that they can use to recreate the constellations of Cygnus and Scorpius. You will need a block of wood, pins, and loom bands for this project. Have the students stick the pins into the wood in a grid. Then, have them use the constellation cards found in the appendix on p. 111 as a guide to recreate the constellations. You can see how this project works here:

- https://www.schooltimesnippets.com/2014/06/constellations-activity-geoboard.html

To-Do List

Essential

- ☐ Read the appropriate reading assignment.
- ☐ Complete the lapbook or notebook assignments.
- ☐ Label the stars.
- ☐ Do the "Planetarium" activity.

Optional

- ☐ Get one or more of the library books to read.
- ☐ Fill out a lab report sheet for one of the activities.
- ☐ Do the "Geoboard Constellations" activity.

Lesson 6: Fall Constellations

Information

Reading Assignments

- **All Students** – *Glow in the Dark Constellations* p. 26 Pegasus, p. 28 Andromeda
- **(Optional) All Students** – *The Stars and Their Stories* pp. 57-76 The Royal Family

(Optional) Books from the Library

- *The Heavenly Zoo: Legends and Tales of the Stars* by Alison Lurie

Notebooking

Writing Instructions

- **Lapbook** – Have the students complete the Fall Constellations tab-book on LT pp. 48. Have them cut and color the cover and the pages for Pegasus and Andromeda. Next, have the students tell you what they have learned about the two constellations and write this for them on the respective pages. Then, staple the pages together and glue the mini-book into the lapbook.
- **Notebook** – Have the students dictate, copy, or write one to four sentences about Pegasus and Andromeda on the Fall Constellations notebooking page on NB p. 33.

Star Labeling

Have the students label the following key stars on the constellations in their lapbook or notebook:

- The Great Square (*Pegasus*)
- Alpheratz (*Andromeda*)

Hands-on Science

Coordinating Activities

- **Planetarium** – Have the students make a homemade planetarium to view the constellations. You will need the planetarium from last week, foil, a pin, a Sharpie, and the constellation pictures for Pegasus and Andromeda on p. 112. Begin by cutting the foil into two squares so that each square will fit around the end of the toilet paper tube and have enough overlap to be secured by the rubber band. Have the students use a pin and the constellation picture from the appendix to create two foil constellations, one for Pegasus and one for Andromeda. Have them use a Sharpie to write the name of the constellation in the bottom corner of each piece of foil. Once they have the two constellations, have the students view each one by placing the foil template at one end of the toilet paper tube so that the constellation is centered in the middle of the tube. Then, secure the foil in place using the rubber band. Next, set the tube on a desk, place the flashlight in the other end of the tube, and aim the foil end of the tube toward a wall. Finally, turn off the lights, and turn on the flashlight to view the constellations. (**Note** – *If you want to take this a step further, have the students make constellation pages for Pegasus and Andromeda.*)

✂ **(Optional) Constellation Party** – Have a constellation party with the students. You can make constellation cupcakes from the recipe found at:

- http://www.piecesbypolly.com/2012/02/easy-constellation-cupcakes-space-party.html

You can also play the Which Constellation Is It? game. For this, use the constellation cards on pp. 110-112 of the appendix and cut the name off each card. Have the students lay out the constellation pictures and then try to match the names with the pictures.

Review Sheet

Stars Review Sheet Answers

1. C, B, D, A
2. Gas
3. C, A, D, B
4. Constellation

To-Do List

Essential

☐ Read the appropriate reading assignment.

☐ Complete the lapbook or notebook assignments.

☐ Label the stars.

☐ Do the "Planetarium" activity.

Optional

☐ Get one or more of the library books to read.

☐ Fill out a lab report sheet for one of the activities.

☐ Complete the Stars Review Sheet. (pp. 128-129)

☐ Have a Constellation Party.

SCIENCE CHUNKS

Matter Unit

Lesson 1: States of Matter

Information

Reading Assignments

- **Younger Students** - *Basher Science Chemistry* p. 8 Solid, p. 10 Liquid, p. 12 Gas
- **Older Students** - *Usborne Science Encyclopedia* pp. 16-17 Solids, Liquids, and Gases
  (**Note** - There is also a fourth state of matter, plasma, which is a bit more abstract. It consists of atoms that have been split by high heat or electricity. Since it is difficult for students to be able to visualize plasma, we have left this information out for their first look at the different states of matter.)

(Optional) Books from the Library

- *What Is the World Made Of? All About Solids, Liquids, and Gases (Let's-Read-and-Find Out About Science)* by Kathleen Weidner Zoehfeld and Paul Meisel
- *Solids, Liquids, And Gases (Rookie Read-About Science)* by Ginger Garrett
- *States of Matter: A Question and Answer Book* by Fiona Bayrock and Anne McMullen

Notebooking

Vocabulary

Go over the following words with your students. Then, have them create flashcards or copy the definitions into the glossary.

- **States of Matter** – The different forms in which a substance can exist: solid, liquid, and gas. (Flashcard LT p. 60; Glossary NB p. 65)
- **Volume** – The space occupied by matter. (Flashcard LT p. 60; Glossary NB p. 67)

Writing Instructions

- **Lapbook** – Have the students begin the Matter lapbook by cutting out and coloring the cover on LT p. 53.
- **Lapbook** – Have the students complete the States of Matter Cut-flap Book on LT p. 55. Have them cut out and fold the template. Have the students color the pictures on the cover. Then, have the students write the definition for solid, liquid, and gas on the inside. Finally, glue the flapbook into the lapbook.
- **Lapbook** – Have the students cut out and glue the vocabulary pocket on LT p. 59 into their lapbook.
- **Lapbook** – Have the students add the "States of Matter" poem to the lapbook. Have them cut out and color the poem sheet found on LT p. 54. Once they are done, have them glue the sheet into the lapbook. (**Note** - You can have the students memorize this poem as you work through this unit.)

Notebook – Have the students dictate, copy, or write three to five sentences about solids, liquids, and gases on the states of matter notebooking page on NB p. 36.

Hands-on Science

Coordinating Activities

Play with Matter – You will need three balloons, water, and ice. Fill one balloon with ice, one balloon with water, and one with gas (air). Let the students explore each of the balloons and observe the differences. As they make their observations, ask them questions like:

- **?** Which one floats the best?
- **?** Which one is easiest to control?
- **?** Which one has the most interesting shape?

(Optional) States of Matter Float – Make a States of Matter Ice-cream Float with your students! You will need ice cream, root beer (or other type of soda), and a glass. Have the students add the solid, ice cream, first. Then, pour over the liquid, root beer, and watch the gas bubbles appear!

To-Do List

Essential

- ☐ Read the appropriate reading assignment.
- ☐ Define states of matter and volume.
- ☐ Complete the lapbook or notebook assignments.
- ☐ Do the "Play with Matter" activity.

Optional

- ☐ Get one or more of the library books to read.
- ☐ Fill out a lab report sheet (p. 101) for one of the activities.
- ☐ Do the "States of Matter Float" activity.

Lesson 2: Changes in State

Information

Reading Assignments

- **Younger Students** - *Basher Science Chemistry* p. 14, Melting Point, p. 15 Boiling Point **(Note** - If you choose this resource, you will need to discuss freezing and condensing with your students.)
- **Older Students** - *Usborne Science Encyclopedia* pp. 18-19 Changes in States

(Optional) Books from the Library

- *How Water Changes (Weekly Reader: Science)* by Jim Mezzanotte
- *Solids (States of Matter)* by Jim Mezzanotte
- *Liquids (States of Matter)* by Jim Mezzanotte
- *Gases (States of Matter)* Jim Mezzanotte

Notebooking

Vocabulary

Go over the following words with your students. Then, have them create flashcards or copy the definitions into the glossary.

- **Physical Change** – A change that occurs in which no new substances are made. (Flashcard LT p. 61; Glossary NB p. 63)
- **Sublimation** – A change from solid to gas without going through liquid form. (Flashcard LT p. 61; Glossary NB p. 66)

Writing Instructions

- **Lapbook** – Have the students complete the Changes in State Arrow-book on LT p. 56. Have the students cut out and fold the template. Have them color the picture on the cover. Then, have the students write several sentences about what they have learned. Finally, glue the flap-book into the lapbook.
- **Notebook** – Have the students dictate, copy, or write three to five sentences about melting, boiling, freezing, and condensing on the changes on state notebooking page on NB p. 37.

Hands-on Science

Coordinating Activities

- **Freezy Meltdown** – You will need an orange juice and a cup. Pour a bit of juice in a cup for each of the students and have them take a sip of the juice. Have the students place the remaining juice in the freezer. After it has hardened (about an hour or two), take the cup out and have the students observe the changes that have occurred. Let the cup sit on the counter in a place where it won't be disturbed. After the juice has melted (about an hour or two), have the students take a sip of the juice to see if it tastes the same as it did during step 1.

✂ **(Optional) Changes in State** – Have the students set several ice cubes on a plate and set the plate out in the hot sun. Check the plate every five minutes to observe what happens. The students should see the solid water (ice) melts into liquid water, which then evaporates into a gas and disappears into the air.

To-Do List

Essential

- ☐ Read the appropriate reading assignment.
- ☐ Define physical change and sublimation.
- ☐ Complete the lapbook or notebook assignments.
- ☐ Do the "Freezy Meltdown" activity.

Optional

- ☐ Get one or more of the library books to read.
- ☐ Fill out a lab report sheet for one of the activities.
- ☐ Do the "Changes in State" activity.

Lesson 3: Liquid Behavior

Information

Reading Assignments

- **Younger Students** - *Basher Science Chemistry* p. 10 Liquid
  (Note - There are no new pages in this resource for this week. If you do not wish to re-read the section on liquids, you will need to choose one of the additional books from your library or read from the Usborne Science Encyclopedia for this week.)
- **Older Students** - *Usborne Science Encyclopedia* pp. 20-21 How Liquids Behave

(Optional) Books from the Library

- *What Is a Liquid? (First Step Nonfiction, States of Matter)* by Jennifer Boothroyd
- *How Do You Measure Liquids? (A+ Books: Measure It!)* by Thomas K. Adamson
- *Saving Water: The Water Cycle (Do It Yourself)* by Buffy Silverman
- *Why Do Puddles Disappear?: Noticing Forms of Water* by Martha E. H. Rustad and Christine M. Schneider

Notebooking

Vocabulary

Go over the following words with your students. Then, have them create flashcards or copy the definitions into the glossary.

- **Evaporation** – The process by which the surface molecules of a liquid escape into a vapor. (Flashcard LT p. 29; Glossary NB p. 60)
- **Surface Tension** – A force that pulls together molecules on the surface of a liquid. (Flashcard LT p. 29; Glossary NB p. 66)

Writing Instructions

- **Lapbook** – Have the students complete the Liquid Behavior mini-book on LT p. 57. Have them cut out the mini-book and color the cover. Next, have the students tell you what they have learned about how liquids behave. Then, have them fold the booklet, glue on the cover, and place it into the lapbook.
- **Notebook** – Have the students dictate, copy, or write three to five sentences about how liquids behave, evaporation, and surface tension on the liquid behavior notebooking page on NB p. 38.

Hands-on Science

Coordinating Activities

- **Surface Tension** – You will need a shallow bowl, liquid soap, and paper clip. Fill the bowl with a thin layer of water and gentle set the paper clip on the top of the water, so that it floats. (The paper clip is being held up by the surface tension of the water.) Then, add a drop of the liquid soap near the paper clip and watch what happens. (The soap will break the surface tension of the water and the paper clip will fall to the bottom of the bowl.)

✂ **(Optional) Evaporation** – M ix together a quarter cup of warm water and a tablespoon of salt. Pour the mixture into a shallow dish and set the dish out in a sunny place. Check the dish every hour or so to observe what happens.

To-Do List

Essential

- ☐ Read the appropriate reading assignment.
- ☐ Define evaporation and surface tension.
- ☐ Complete the lapbook or notebook assignments.
- ☐ Do the "Surface Tension" activity.

Optional

- ☐ Get one or more of the library books to read.
- ☐ Fill out a lab report sheet for one of the activities.
- ☐ Do the "Evaporation" activity.

Lesson 4: Gas Behavior

Information

Reading Assignments

- **Younger Students** - *Basher Science Chemistry* p. 12 Gas, p. 14 Brownian Motion
- **Older Students** - *Usborne Science Encyclopedia* p. 16 Section on Brownian Motion, pp. 22-23 How Gases Behave

(Optional) Books from the Library

- *What Is a Gas? (First Step Nonfiction)* by Jennifer Boothroyd
- *It's a Gas!* by Ruth Griffin, Margaret Griffin and Pat Cupples

Notebooking

Vocabulary

Go over the following words with your students. Then, have them create flashcards or copy the definitions into the glossary.

- **Diffusion** – The spreading out of a gas to fill the available space. (Flashcard LT p. 63; Glossary NB p. 59)

Writing Instructions

- **Lapbook** – Have the students complete the Gas Behavior mini-book on LT p. 58. Have them cut out the mini-book and color the cover. Next, have the students tell you what they have learned about how gases behave. Then, have them fold the booklet, glue on the cover, and place it into the lapbook.
- **Notebook** – Have the students dictate, copy, or write three to five sentences about Brownian motion, diffusion, and pressure and temperature on the gas behavior notebooking page on NB p. 39

Hands-on Science

Coordinating Activities

- **Diffusion of Gases** – You will need vanilla extract, an eyedropper, and a cotton ball. Place a few drops of vanilla extract on the cotton ball and then wait for a few minutes. Have the students raise their hands as soon as they can smell the vanilla. (The students are able to smell the vanilla as the molecules of vanilla gas move, or diffuse, through the air to reach their noses.)
- **(Optional) Brownian Motion Painting** – You will need a small ball (marble or rubber bouncy ball), paint, paper, and a tray. Place the paper in the tray. Then, have the students dip the ball into the paint and place it in the tray. Have them pick up the tray and shake it gently

with random motions. The ball will move around and create random paths similar to the Brownian motion of a molecule in a liquid or gas.

Review Sheet

Matter Review Sheet Answers

1. B, C, A
2. True
3. Melts, Boils
4. Freezes, Condenses
5. Shape, Volume
6. True
7. High, Low
8. Pressure

To-Do List

Essential

☐ Read the appropriate reading assignment.

☐ Define diffusion.

☐ Complete the lapbook or notebook assignments.

☐ Do the "Diffusion of Gases" activity.

Optional

☐ Get one or more of the library books to read.

☐ Fill out a lab report sheet for one of the activities.

☐ Complete the Matter Review Sheet. (pp. 130-131)

SCIENCE CHUNKS

Acids and Bases Unit

Lesson 1: Acids and Bases

Information

Reading Assignments

- **Younger Students** - *Basher Science Chemistry* p. 46 Acid, p. 48 Base
- **Older Students** - *Usborne Science Encyclopedia pp.* 84-85 Acids and Bases

(Optional) Books from the Library

- *Acids and Bases (Why Chemistry Matters)* by Lynnette Brent
- *Acids and Bases (Material Matters)* by Carol Baldwin
- *Acids and Bases (Chemicals in Action)* by Chris Oxlade

Notebooking

Vocabulary

Go over the following words with your students. Then, have them create flashcards or copy the definitions into the glossary.

- **Acid** – A chemical that dissolves in water and can neutralize a base. Weak acids taste sour. (Flashcard LT p. 72; Glossary NB p. 56)
- **Base** – A chemical that dissolves in water and can neutralize an acid. Weak bases taste bitter. (Flashcard LT p. 72; Glossary NB p. 57)

Writing Instructions

- **Lapbook** – Have the students begin the Acids lapbook by cutting out and coloring the cover on LT p. 66.
- **Lapbook** – Have the students complete the Acids and Bases tab-book-book on LT p. 68. Have them cut out and color the pictures on the cover. Then, have the students write what they have learned about acids and bases on their respective tabs. Have them staple the mini-book together and glue it into the lapbook.
- **Lapbook** – Have the students cut out and glue the vocabulary pocket on LT p. 71 into their lapbook.
- **Lapbook** – Have the students cut out and color the "Acids and Bases" poem on LT p. 67. Once they have finished, have them glue the poem into the lapbook. (**Note** - You can have the students memorize this poem as you work through this unit.)
- **Notebook** – Have the students dictate, copy, or write one to four sentences on acids and bases on NB p. 42.

Hands-on Science

Coordinating Activities

✂ **Crazy Colors** – You will need a head of red cabbage, knife, pot, water, strainer, a large plastic container, two clear glasses, ammonia, and white vinegar. Chop up a third of the red cabbage head. Place it in the pot and cover it with water. Bring the water to a boil and let it boil for 10 to 15 minutes—basically until the water is a deep purple color. Let cool the red cabbage juice before straining about a cup of the liquid into the plastic container. Next, have the students add about a quarter of a cup of the red cabbage juice to each of the clear glasses. As they watch, add a tablespoon of ammonia to one of the glasses. (You should see the liquid go from deep purple to bright neon green! If you don't, add another tablespoon ammonia.) Then, add a tablespoon of vinegar to the other glass. (You should see the liquid go from deep purple to bright pink! If you don't, add another tablespoon vinegar.) Red cabbage juice acts as an indicator of pH, which is a fancy way of saying that it changes color. In the presence of an acid, the juice can turn pink. In the presence of a base, it can turn bright green. Ammonia is a base, vinegar is an acid. (**Note** – You will need red cabbage juice again in this unit. So be sure to save what you have left.)

✂ **(Optional) Acids and Bases Poster** – Over this unit, the students will create two posters with the acids and bases they encounter in real life. For example, this week, they could add lemonade to the acid poster and baking soda to the base poster. They can simply write the names or they can use pictures of the acids and bases.

To-Do List

Essential

- ☐ Read the appropriate reading assignment.
- ☐ Define acid and base.
- ☐ Complete the lapbook or notebook assignments.
- ☐ Do the "Crazy Colors" activity.

Optional

- ☐ Get one or more of the library books to read.
- ☐ Fill out a lab report sheet (p. 101) for one of the activities.
- ☐ Do the "Acids and Bases Poster" activity.

Lesson 2: pH

Information

Reading Assignments

- **Younger Students** - *Basher Science Chemistry* p. 50 pH, p. 52 Indicator
- **Older Students** - *Usborne Science Encyclopedia* pp. 86-87 Acids and Bases, part 2

(Optional) Books from the Library

- There are no additional books on the market for these topics. Instead, you can watch the following videos on pH:
  - Importance of pH in everyday life: https://www.youtube.com/watch?v=x-nI-3Ws7nxQ
  - What is the pH scale? https://www.youtube.com/watch?v=3U9n4BV2618

Notebooking

Vocabulary

Go over the following words with your students. Then, have them create flashcards or copy the definitions into the glossary.

- **Indicator** – A substance that changes color in the presence of an acid or base. (Flashcard LT p. 73; Glossary NB p. 61)
- **pH** – A scale from 0 to 14 used to measure the strength of acids and bases. (Flashcard LT p. 73; Glossary NB p. 63)

Writing Instructions

- **Lapbook** – Have the students complete the pH Scale Sheet on LT p. 69. Have the students cut out and write whether the section is acidic, basic, or neutral. Then, have the students color the acid range from dark (at 0) to light (before 7) red. Have them color the base section from dark (at 14) to light (before 7) blue. Have them leave the neutral section white. Finally, glue the sheet into the lapbook.
- **Notebook** – Have the students dictate, copy, or write one to four sentences on pH and indicators on NB p. 43.

Hands-on Science

Coordinating Activities

- **pH Test** – Have the students create their own pH scale for cabbage juice indicator. You can get directions and view an example in the following post:
  - http://elementalblogging.com/science-corner-kitchen-acid-test/

✂ **(Optional) Indicator** – Have the students use another indicator to test the acids and bases they know about so far. Grape juice is another option for an indicator. It turns reddish-purple in the presence of an acid and purplish-black in the presence of a base.

To-Do List

Essential

- ☐ Read the appropriate reading assignment.
- ☐ Define indicator and pH.
- ☐ Complete the lapbook or notebook assignments.
- ☐ Do the "pH" activity.

Optional

- ☐ Get one or more of the library books to read.
- ☐ Fill out a lab report sheet for one of the activities.
- ☐ Do the "Indicator" activity.

Lesson 3: Salts

Information

Reading Assignments

- **Younger Students** - *Basher Science Chemistry* p. 30 Ions, "Neutralization" on p. 37 of the appendix
- **Older Students** - *Usborne Science Encyclopedia* pp. 88-90 Salts

(Optional) Books from the Library

- *From Sea to Salt (Start to Finish, Second Series)* by Lisa Owings
- *The Story of Salt* by Mark Kurlansky and S. D. Schindler

Notebooking

Vocabulary

Go over the following words with your students. Then, have them create flashcards or copy the definitions into the glossary.

- **Neutralization** – A reaction where one substance fully or partly cancels out another. (Flashcard LT p. 74; Glossary NB p. 62)
- **Salt** – An ionic compound that when dissolved in water makes positive and negative. (Flashcard LT p. 74; Glossary NB p. 64)

Writing Instructions

- **Lapbook** – Have the students begin the Neutralization sheet on LT p. 70. Have them cut out the sheet and label the pictures "acid", "base", "water", "salt". Then, have them glue the sheet into the lapbook.
- **Notebook** – Have the students dictate, copy, or write one to four sentences on neutralization and salts on NB p. 44.

Hands-on Science

Coordinating Activities

- **Neutralize It** – You will need vinegar, baking soda, water, cabbage juice, coffee filter, 2 clear cups, and eyedropper. First, make the cabbage paper using the cabbage juice and coffee filter. Dip the coffee filter in the cabbage juice and then set it aside to dry. Once the coffee filter dries, cut it into strips to use as your cabbage paper. Second, pour a quarter of a cup of vinegar in the glass and set it in the sink. Dip a strip of your cabbage paper into the vinegar to test to see if it is an acid. (It should turn a pinkish color, indicating the presence of an acid.) Third, in another cup, mix 1 tablespoon of baking soda with a half of a cup of warm water. Stir until dissolved then dip a strip of cabbage paper into the vinegar to test to see if it is a base. (It should turn into a bluish color, indicating the presenf of a base.) Fourth, have the students add about a teaspoon of the cabbage juice to the cup in the sink, which should turn the vinegar solution to a pinkish color. Finally, use the eyedropper to add the baking soda solution to the vinegar until the color changes to purple, indicating a neutral solution. (This activity explains the occurence of neutralization when an acid and a base are mixed in the right proportions. After some carbon dioxide is given off in the reaction, the solution now contains only water and a salt (sodium acetate) made from the acid and the base.)

✂ **(Optional) Neutralization** – Let the students have some fun with acids and bases by making some drinkable fizzing lemonade! The directions for this project can be found at the following blog post:

- https://www.learnwithplayathome.com/2014/09/how-to-make-fizzing-lemon-ade-edible.html

Review Sheet

Acids and Bases Review Sheet Answers

1. True
2. False *(A base dissolves in water and can taste bitter.)*
3. Power of hydrogen
4. pH 2 - acid, pH 7 - neutral, ph 10 - base
5. Water, salt
6. Ions

To-Do List

Essential

- ☐ Read the appropriate reading assignment.
- ☐ Define neutralization and salt.
- ☐ Complete the lapbook or notebook assignments.
- ☐ Do the "Neutralize It" activity.

Optional

- ☐ Get one or more of the library books to read.
- ☐ Fill out a lab report sheet for one of the activities.
- ☐ Complete the Acids and Bases Review Sheet. (p. 132)

SCIENCE CHUNKS

Pasteur Unit

Lesson 1: Louis Pasteur

Information

Reading Assignments

📖 **All Students** – *Pasteur's Fight Against Microbes* by Beverly Birch (**Note**—You can read this book all in one sitting or break it up into two sessions by stopping at the page that ends with, "A war had been fought and won.")

Discussion Questions

When you are done, you can ask the students the following questions:

- What was Louis Pasteur's job? (Louis Pasteur taught the boys of Lille.)
- What problem was Monsieur Pasteur presented with at the sugarbeet factory? (Monsieur Bigo showed Monsieur Pasteur how the sugarbeet juice was turning sour, slimy, and gray instead of fermenting.)
- Did Pasteur know how to fix the sugarbeet problem? How did he begin? (Pasteur had no idea how to fix the sugarbeet problem, but he began by taking a look at a sample of the good batch of fermented sugarbeet juice and the foul batch of fermented sugarbeet juice.)
- What did Pasteur find when he looked at the two samples under a microscope? And what did he conclude? (In the good batch sample, he saw tiny, swimming things that he identified as yeast globules. Pasteur also saw the yeasts swell to form buds that broke away and repeated the process. He concluded that the yeasts in the good batch were alive, feeding on the beet juice to produce alcohol. In the foul batch sample, he saw no yellow, yeast globules. When he looked at the gray specks floating on the foul liquid, he found a lot of tiny black rods, which were also alive. He concluded that the black rods had taken over and stopped the yeast from making alcohol in the foul batch.)
- Did Pasteur immediately share what he found? (No, he took more samples to see if he saw the same thing in each one.)
- After he solved the mystery at the sugarbeet factory, did Pasteur stop thinking about the black rods he found? (No, he continued to do several experiments until he could reproduce the gray stuff with the tiny, black rods in his lab.)
- What idea about microbes did Pasteur show to be different? (Scientists in Pasteur's time thought that microbes were on things that were rotting because the rotting caused the microbes to come to life. Instead, Pasteur showed that microbes were actually the cause of the rotting.)
- Where did Pasteur find microbes? (Pasteur found that microbes were in the air on the dust particles that floated around. He found more microbes in the city air than in the country or mountain air.)
- How did Pasteur's discovery about microbes change our world? (Scientists were able to look at

where microbes came from, learn how to stop them, and stop the spread of many diseases.)

- **(Older Students)** Why do you think Pasteur was willing to tackle the problem, even though he had no idea what was going on at the sugarbeet factory? Do you think this is important for scientists to do? (Answers will vary, but students can include something about how Pasteur wanted to solve the mystery.)
- **(Older Students)** Do you think it was important for Pasteur to repeat his experiments multiple times? (Answers will vary, but students could say that yes it was because it showed that he had found the answer instead of a one-time fluke.)
- **(Older Students)** After reading about Pasteur's major discovery, what characteristics do you think are important for scientists to have? (Answers will vary, but students can include characteristics like curiosity and perseverance.)

(Optional) Books from the Library

- *Louis Pasteur and Pasteurization* by Jennifer Lee Fandel
- *Germ Hunter: A Story about Louis Pasteur* by Elaine Marie Alphin and Elaine Verstraete
- *Louis Pasteur: Founder of Modern Medicine* by John Hudson Tiner

Notebooking

Vocabulary

Go over the following words with your students. Then, have them create flashcards or copy the definitions into the glossary.

- **Fermentation** – A chemical reaction that breaks down sugar into carbon dioxide and alcohol. (Flashcard LT p. 81; Glossary NB p. 60)
- **Microbes** – A microscopic organism that can cause disease. (Flashcard LT p. 82; Glossary NB p. 61)
- **Yeast** – A microscopic, single-celled fungus. It is used in fermentation. (Flashcard LT p. 82; Glossary NB p. 68)

Writing Instructions

- **Lapbook** – Have the students complete the Louis Pasteur tab-book on LT pp. 78-79. Have them cut out the mini-book pages and color the cover. Next, have the students tell you what they have learned about Louis Pasteur and his work on the respective tabs. Then, have them staple the pages of the booklet together and glue it into the lapbook.
- **Lapbook** – Have the students complete the Pasteur timeline sheet on LT p. 80. Have them cut out the timeline sheet and color the pictures. Next, have the students add one to three of the significant dates (see below) on the timeline.
- **Lapbook** – Have the students cut out and glue the vocabulary pocket on LT p. 81 into their lapbook.
- **Notebook** – Have the students add the following dates to the Pasteur timeline notebooking page on NB p. 46-48.

- 1822 – Louis Pasteur is born in France.
- 1847 – Pasteur goes to college and studies the sciences.
- 1855 – Pasteur finds microbes in foul-smelling sugarbeet juice.
- 1864 – Louis Pasteur finds that microbes are in the air. He also develops a germ theory showing that microbes are responsible for getting people sick.
- 1868 – Pasteur has a small stroke that leaves him partially paralyzed, but still able to work.
- 1879 – Louis Pasteur accidentally discovers vaccines.
- 1895 – Louise Pasteur dies.

Notebook – Have the students dictate, copy, or write information about Louis Pasteur and his work on the Pasteur notebook pages on NB p. 46-48.

Hands-on Science

Coordinating Activity

Yeast Fermentation – Have the students watch how yeast ferments. You will need a packet of yeast, warm water, sugar, and a clear container. Add a teaspoon of sugar and a packet of yeast to the container. Then add about a cup of warm water and gently stir the mixture. Observe what happens in the container over the next 20 minutes. (The students should see bubbles form and these bubbles create a foam on top of the water. They should also start to smell a sweet scent similar to the smell of bread baking.)

(Optional) Making Bread – Have the students use yeast to make bread. You can watch the following video on the chemistry of bread-making first:

https://www.youtube.com/watch?v=qylxpwNhFYI

Then, use the directions on the back of the yeast packet to make rolls.

Review Sheet

Pasteur Review Sheet Answers

1. C, B, A
2. True
3. False (Louis Pasteur did his experiments over and over to confirm that his results remained the same.)
4. True
5. True
6. Answers will vary.

To-Do List

Essential

- ☐ Read the appropriate reading assignment.
- ☐ Define fermentation, microbes, and yeast.
- ☐ Do the "Yeast Fermentation" activity.
- ☐ Complete the lapbook or notebook assignments.

Optional

- ☐ Get one or more of the library books to read.
- ☐ Do the "Making Bread" activity.
- ☐ Complete the Pasteur Review Sheet on pp. 133.

SCIENCE CHUNKS

Energy Unit

Lesson 1: Energy Basics

Information

Reading Assignments

- **Younger Students** - *DK Children's Encyclopedia* pp. 88-89 Energy
- **Older Students** - *Usborne Science Encyclopedia* pp. 106-110 Energy

(Optional) Books from the Library

- *Energy (Science Readers)* by Suzanne I. Barchers
- *Energy (True Books: Physical Science)* by Matt Mullins
- *Energy Makes Things Happen (Let's-Read-and-Find-Out Science 2)* by Kimberly Brubaker Bradley and Paul Meisel

Notebooking

Vocabulary

Go over the following words with your students. Then, have them create flashcards or copy the definitions into the glossary.

- **Energy** – The ability to do work. (Flashcard LT p. 93; Glossary NB p. 59)
- **Energy Chain** – A way of showing how energy changes into different forms. (Flashcard LT p. 94; Glossary NB p. 59)

Writing Instructions

- **Lapbook** – Have the students cut out the pages for the Energy Tab-book on LT p. 86 and color the pictures. Then, have the students add a sentence about potential energy on the potential page and a sentence about kinetic energy on the kinetic page. Assemble the tab-book and staple it together on the dashed lines. Finally, have the students glue the mini-book into the lapbook.
- **Lapbook** – Have the students complete the Energy Law Sheet on LT p. 87. Have them cut out the sheet and copy the Law of Conservation of Energy in the space provided.

  <u>Law of Conservation of Energy</u>
  Energy can neither be created nor destroyed.

  Then, have the students glue the sheet into the lapbook. (**Note** - *You can have the students memorize this law as you work through this unit.*)
- **Notebook** – Have the students dictate, copy, or write three to five sentences about the basics of energy on the energy notebooking page on NB p. 50.

Hands-on Science

Coordinating Activity

- **Energy Race** – Have the students compete to see who can transfer the most energy to their rubber band. You will need several people, a rubber band for each person, and a measuring tape. Draw a line at one end of a room or outside. Give each player a rubber band and have them stand on the line. Call out "potential," at which point the players will stretch their rubber bands. Then, call out "kinetic," at which point the players will let go. Measure the distance each rubber band has traveled. The player whose rubber band has traveled the farthest wins the race! *(You can also have several trials and add up the distances to see who is the energy winner.)*
- **(Optional) Energy Boat** – Have the students do the "See for yourself " activity on p. 107 of the *Usborne Science Encyclopedia*. You will need a matchbox, a cardboard, two used matches, and a rubber band for this activity.

To-Do List

Essential

- ☐ Read the appropriate reading assignment.
- ☐ Define energy and energy chain.
- ☐ Complete the lapbook or notebook assignments.
- ☐ Do the "Energy Race" activity.

Optional

- ☐ Get one or more of the library books to read.
- ☐ Fill out a lab report sheet (p. 101) for one of the activities.
- ☐ Do the "Energy Boat" activity.

Lesson 2: Energy Resources

Information

Reading Assignments

- **Younger Students** - *DK Children's Encyclopedia* p. 110 Fossil Fuels
  You can also watch the following video to learn more about wind energy:
  - https://www.youtube.com/watch?v=SQpbTTGe_gk
- **Older Students** - *Usborne Science Encyclopedia* pp. 108-109 Energy Resources

(Optional) Books from the Library

- *Sun Power: A Book about Renewable Energy (Earth Matters)* by Esther Porter
- *Wind Energy: Blown Away! (Powering Our World)* by Amy S. Hansen
- *Solar Energy: Running on Sunshine (Powering Our World)* by Amy S. Hansen
- *Energy from the Sun (Rookie Read-About Science)* by Allan Fowler
- *Biomass Power (Let's Discuss Energy Resources)* by Richard Spilsbury
- *Fossil Fuel Power (Let's Discuss Energy Resources)* by Richard Spilsbury

Notebooking

Vocabulary

Go over the following words with your students. Then, have them create flashcards or copy the definitions into the glossary.

- **Solar Energy** – Energy from the sun. (Flashcard LT p. 94; Glossary NB p. 65)
- **Wind Energy** – Energy from the wind. (Flashcard LT p. 95; Glossary NB p. 68)

Writing Instructions

- **Lapbook** – Have the students cut out the wheels for the Energy Resources Wheel-Book on LT p. 87. On one half of the wheel, have the students add the definition of renewable resources along with several examples of renewable energy. On the other half of the wheel, have the students add the definition of nonrenewable resources along with several examples of nonrenewable energy. Then, use a brad to attach the two pages together so that the cover spins to reveal one half at a time. Finally, have them glue the mini-book into the lapbook.
  - **Non-renewable energy resources** – Energy sources that exist in limited quantities, such as oil and coal.
  - **Renewable energy resources** – Energy sources that can be replaced in a relatively short period of time, such as wind and solar.
- **Notebook** – Have the students dictate, copy, or write three to five sentences on energy resources on the energy resources notebooking page on NB p. 51.

Hands-on Science

Coordinating Activity

- **Solar Oven** – Have the students build their own solar oven out of an old pizza box. You will need scissors, plastic wrap, aluminum foil, tape, an old pizza box, black construction paper, glass or metal pie plate, and a few marshmallows or a piece of buttered toast. Use the scissors to make a flap out of the top of the pizza box by cutting three sides, leaving about an inch away from the sides. Then, fold the flap back and cover the flap with aluminum foil. Next, cover the open hole in the top with plastic wrap. After that, line the bottom of the box with black construction paper. Now, place your marshmallows or piece of buttered toast on the pie plate, set the plate inside the oven, and take the oven outside. Finally, position the box and flap so that the sun's rays are directed towards the plastic-wrap covered opening. Check the oven every 10 to 15 minutes to see when your food is done. Use a hot mitt to remove the food and enjoy your solar-powered meal!
- **(Optional) Renewable Heat** – Have the students do the "See for yourself " activity on p. 109 of the *Usborne Science Encyclopedia*. You will need a hose and a cork for this activity.

To-Do List

Essential

- ☐ Read the appropriate reading assignment.
- ☐ Define solar energy and wind energy.
- ☐ Complete the lapbook or notebook assignments.
- ☐ Do the "Solar Oven" activity.

Optional

- ☐ Get one or more of the library books to read.
- ☐ Fill out a lab report sheet for one of the activities.
- ☐ Do the "Renewable Heat" activity.

Lesson 3: Nuclear Energy

Information

Reading Assignments

- **Younger Students** - No pages scheduled for younger students. Instead, have the students watch the following video to learn more about nuclear power:
  - https://www.youtube.com/watch?v=FNNKhE1FNNM
- **Older Students** - *Usborne Science Encyclopedia* pp. 116-117 Nuclear Power

(Optional) Books from the Library

- *Nuclear Energy: Amazing Atoms (Powering Our World)* by Amy S. Hansen
- *Nuclear Energy (Saving the Planet Through Green Energy)* by Colin Grady
- *Nuclear Energy (Discovery Channel School Science)* by Michael Burgan and Nancy Cohen
- *Nuclear Meltdowns (True Books)* by Peter Benoit
- *Nuclear Power (Energy for Today)* by Tea Benduhn

Notebooking

Vocabulary

Go over the following words with your students. Then, have them create flashcards or copy the definitions into the glossary.

- **Nuclear Fusion** – The joining of atomic particles to create energy. (Flashcard LT p. 95; Glossary NB p. 63)
- **Nuclear Fission** – The splitting apart of atomic particles to create energy. (Flashcard LT p. 96; Glossary NB p. 63)

Writing Instructions

- **Lapbook** – Have the students cut out the pages for the Nuclear Energy tab-book on LT p. 88 and color the pictures. Then, have the students add a sentence about nuclear fission on the fission page and a sentence about nuclear fusion on the fusion page. Assemble the tab-book and staple it together on the dashed lines. Finally, have the students glue the mini-book into the lapbook.
- **Notebook** – Have the students dictate, copy, or write three to five sentences on fission and fusion on the nuclear energy notebooking page on NB p. 52.

Hands-on Science

Coordinating Activity

- **Radioactive Half-life** – Have the students learn about the half-life of a radioactive element. You will need bite-sized food, such as M&M's, raisins, or cereal puffs, plus a timer. Begin by giving each student 32 pieces of the bite-sized food. After 2 minutes, have them eat 16 pieces. After 2 more minutes, have them eat 8 pieces. After 2 more minutes, have them eat 4 pieces. After 2 more minutes, have them eat 2 pieces. After 2 more minutes, have them eat 1 piece. After 2 more minutes, have them break the 1 piece in half and eat one of the halves. After 2 more minutes, have the students eat any of the remaining crumbs. *(This activity is meant to give a students a mental picture of how a half-life works. The elements used in creating nuclear energy are radioactive, which means that they are unstable and they decay with a half life. The half-life period depends upon the element and how radioactive it is. Two of the most common elements used in generating nuclear power are Uranium, which has a half-life or around 4.5 million years, and Plutonium, which has a half-life of around 24,000 years.)*

(Optional) Nuclear Fusion – Have the students learn see how nuclear fusion works. You will need at least two marshmallows. Have the students take two marshmallows in their hands and press them together really hard until the marshmallows stick together, forming a larger marshmallow ball. (*Nuclear fusion works in much the same way - two small nuclei are joined together to form a larger one. This process takes place at extremely high temperatures, but it releases huge amounts of energy. The stars in our solar system, including our sun, use nuclear fusion to produce light.*)

To-Do List

Essential

- ☐ Read the appropriate reading assignment.
- ☐ Define nuclear fusion and nuclear fission.
- ☐ Complete the lapbook or notebook assignments.
- ☐ Do the "Radioactive Half-life" activity.

Optional

- ☐ Get one or more of the library books to read.
- ☐ Fill out a lab report sheet for one of the activities.
- ☐ Do the "Nuclear Fusion" activity.

Lesson 4: Heat Energy

Information

Reading Assignments

- **Younger Students** - *DK Children's Encyclopedia* p. 252 Temperature
- **Older Students** - *Usborne Science Encyclopedia* pp. 110-111 Heat

(Optional) Books from the Library

- *Energy: Heat, Light, and Fuel (Amazing Science)* by Darlene R. Stille and Sheree Boyd
- *Temperature: Heating Up and Cooling Down (Amazing Science)* by Darlene R. Stille and Sheree Boyd
- *Temperature (Blastoff! Readers: Understanding Weather)* by Kristin Schuetz
- *What Is Temperature? (Weather Close-Up)* by Robin Johnson
- *Temperature (Blastoff! Readers Level 4: First Science)* by Kay Manolis

Notebooking

Vocabulary

Go over the following words with your students. Then, have them create flashcards or copy the definitions into the glossary.

- **Heat** – A form of energy that flows from one place to another because of differences in temperature. (Flashcard LT p. 96; Glossary NB p. 60)
- **Temperature** – A measure of how much heat an object has. (Flashcard LT p. 97; Glossary NB p. 66)

Writing Instructions

- **Lapbook** – Have the students cut out and fold the Heat vs. Temperature Shutterfold Book template on LT p. 89. Have the students color the pictures on the cover. Then, have the students write the definition of heat under the heat flap and the definition of temperature under the temperature flap. Finally, have them glue the mini-book into the lapbook.

- **Notebook** – Have the students dictate, copy, or write three to five sentences on heat and temperature on the heat energy notebooking page on NB p. 53.

Hands-on Science

Coordinating Activity

- **Hot or Cold** – Have the students see if they can tell the difference between hot and cold! You will need three bowls, hot-to-the-touch (but not burning) water, cold water, and room temperature water. Fill the first bowl with the hot-to-the-touch (but not burning) water. Fill the second bowl with the room temperature water. Fill the third bowl with the cold water.

Set the three bowls in order in a row on a table. Have the students place their left hand in the hot-to-the-touch (but not burning) water and their right hand in the cold water. Have them keep their hands there for two minutes. When the time is up, have them place both hands into the middle bowl and tell you what temperature the middle bowl is at. *(The students should see that their left hand feels like the water is ice-cold, while their right hand feels like the water is warm or hot. This is because the hand than was in the hot water loses heat when placed in the room temperature water, making the water feel cool, while the hand that was in the cold water gains heat when placed in the room temperature water, making the water feel warm.)*

(Optional) Homemade Thermometer – Have the students make a thermometer. You will need a small plastic bottle, a clear straw, air-dry clay, food coloring, 2 bowls, water, and a permanent marker. Fill one of the bowls halfway with hot water. Fill the second bowl halfway with ice-cold water. Then, have the students add a few drops of food coloring to the small plastic bottle and fill it to the top with lukewarm water. Insert the straw into the bottle and use the air-dry clay to fix it so that the straw is hanging about halfway into the bottle, forming a tight seal on the top. Have the students use the permanent marker to mark the water level in the straw. Next, place the bottle into the bowl with the hot water and observe what happens. (*The students should see that the water level in the straw rises.*) After two minutes, have the students place the bottle into the bowl with the ice-cold water and observe what happens. (*The students should see that the water level in the straw drop.*)

To-Do List

Essential

- ☐ Read the appropriate reading assignment.
- ☐ Define heat and temperature.
- ☐ Complete the lapbook or notebook assignments.
- ☐ Do the "Hot or Cold" activity.

Optional

- ☐ Get one or more of the library books to read.
- ☐ Fill out a lab report sheet for one of the activities.
- ☐ Do the "Homemade Thermometer" activity.

Lesson 5: Heat Transfer

Information

Reading Assignments

- **Younger Students** - *DK Children's Encyclopedia* p. 157 Materials
  There are no pages scheduled for younger students about heat transfer. Instead, have the students watch the following video:
  - https://www.youtube.com/watch?v=GRY6MpN2QW8
- **Older Students** - *Usborne Science Encyclopedia* pp. 112-113 Heat Transfer

(Optional) Books from the Library

- *How Heat Moves (Science Readers)* by Sharon Coan
- *The Energy That Warms Us: A Look at Heat (Lightning Bolt Books)* by Jennifer Boothroyd

Notebooking

Vocabulary

Go over the following words with your students. Then, have them create flashcards or copy the definitions into the glossary.

- **Conduction** – The transfer of heat through direct contact. (Flashcard LT p. 97; Glossary NB p. 58)
- **Convection** – The transfer of heat through the movement of a liquid and gas. (Flashcard LT p. 98; Glossary NB p. 58)
- **Radiation** – The transfer of heat through indirect contact. (Flashcard LT p. 98; Glossary NB p. 64)

Writing Instructions

- **Lapbook** – Have the students cut out the pages for the Heat Transfer Tab-book on LT p. 92 and color the pictures. Then, have the students write the definition of each of the three different ways to transfer heat (convection, conduction, and radiation) on their respective pages. Once they are done, staple the mini-book together and glue it into the lapbook.
- **Lapbook** – Have the students finish the Energy lapbook on LT p. 92 by cutting out and coloring the cover. Once they are done, have them glue the sheet into the lapbook.
- **Notebook** – Have the students dictate, copy, or write three to five sentences on heat transfer on the heat transfer notebooking page on NB p. 54.

Hands-on Science

Coordinating Activity

- **Water Heater** – Have the students heat up a bowl of water using radiation. You will need a sunny window sill, a bowl, and cool water. Have students fill up a bowl halfway with cool water and then set the bowl on a sunny window sill. Have them check the temperature of the water every thirty minutes and observe the changes. *(The students should see that the water heats up over time.)*

✂ **(Optional) Radiating Heat** – Have the students do the "See for yourself " activity on p. 112 of the *Usborne Science Encyclopedia*. You will need a small feature and a heat source for this activity.

Review Sheet

Energy Review Sheet Answers

1. Potential energy - energy that is stored, Kinetic energy - energy of motion
2. Energy, created, destroyed
3. A renewable source of energy can generate power without being used up. (*Students can also include examples, such as the Sun, wind, or water, for their answer.*)
4. A nonrenewable source of energy can be use only once to generate power. (*Students can also include examples, such as coal, oil, or gas, for their answer.*)
5. False (*Nuclear reactors use controlled fission reactions.*)
6. True
7. Heat is a type of energy that flows from one place to another due to differences in temperature.
8. Thermometer
9. B, A, C
10. False (*A conductor transfer heat quickly, while an insulator transfers heat slowly.*)

To-Do List

Essential

- ☐ Read the appropriate reading assignment.
- ☐ Define conduction, convection, and radiation.
- ☐ Complete the lapbook or notebook assignments.
- ☐ Do the "Water Heater" activity.

Optional

- ☐ Get one or more of the library books to read.
- ☐ Do the "Radiating Heat" activity.
- ☐ Complete the Energy Review Sheet. (pp.134-135)

SCIENCE CHUNKS

Appendix

Activity: ____________________

What I learned

Lab Report: ____________________

Our Tools

Our Method

What it looked like

Our Outcome

Our Insight

Model Joints Directions

Pivot Joint

Supplies Needed

- ✓ A brad
- ✓ Card stock
- ✓ Scissors
- ✓ Hole Punch

Instructions

Cut two rectangles out of the card stock. Then, punch a hole at the end of each of the rectangles. Use the brad to fasten the two together to create a working pivot joint.

Hinge Joint

Supplies Needed

- ✓ 2 Lengths of 1x2 boards
- ✓ ½ Inch hinge
- ✓ Several screws
- ✓ Screwdriver

Instructions

Lay the two boards end to end. Open the hinge and lay it across the point at which the boards meet. Use the screws and screwdriver to attach a side of the hinge to each of the boards. Once it is secure, you can open and shut the boards just like a hinge joint.

(Optional) Ball and Socket Joint

Supplies Needed

- ✓ 1 Length of ½"plastic pipe
- ✓ 1 Length of ¾" plastic pipe
- ✓ 1 ½" Plastic pipe tee
- ✓ 1 ¾" Plastic pipe tee
- ✓ Pipe saw
- ✓ 2 Rubber bands

Instructions

Use the pipe saw to cut the top of the ¾" tee in half so that it forms a cup. Fit the ¾" pipe into the cut ¾" tee. Fit the ½" pipe into the ½" tee. Now fit the ½" tee into the cup of the cut ¾" tee. Use the two rubber bands to secure the two tees together and create a working ball and socket joint.

Inhalation versus Exhalation Worksheet

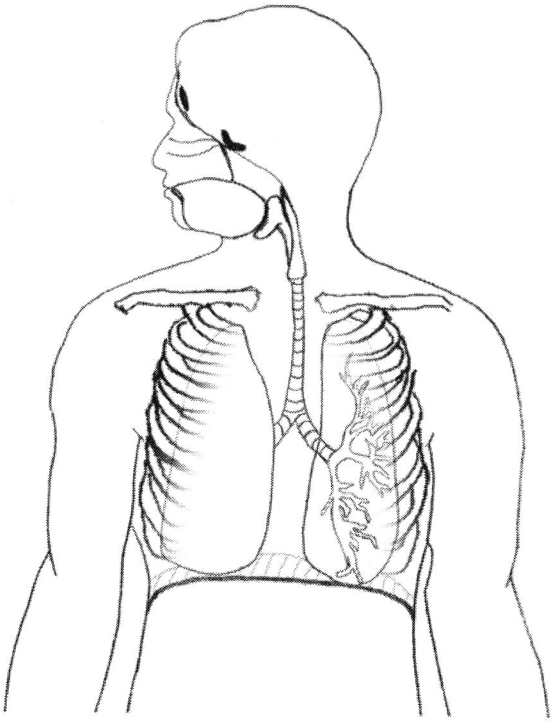

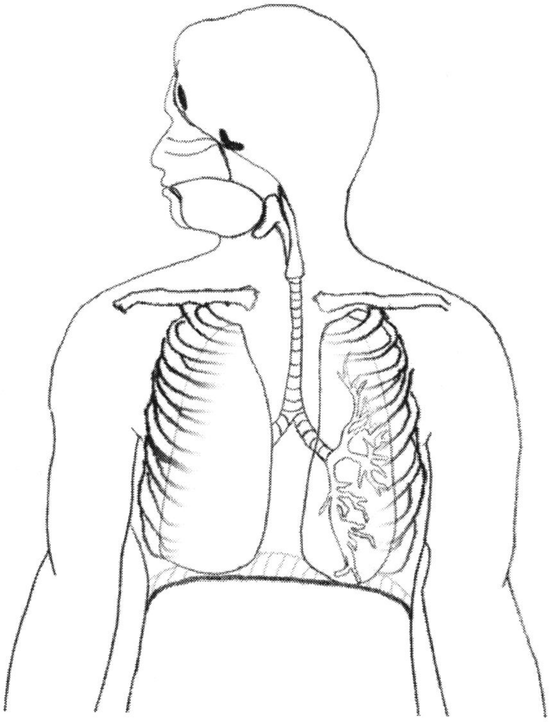

Human Body Pictures

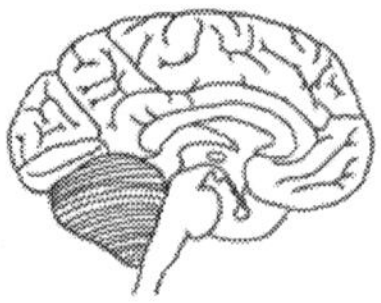

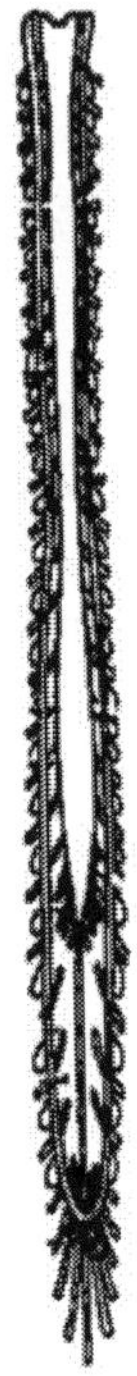

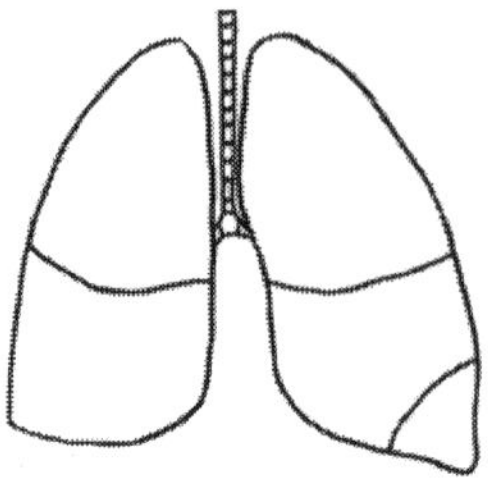

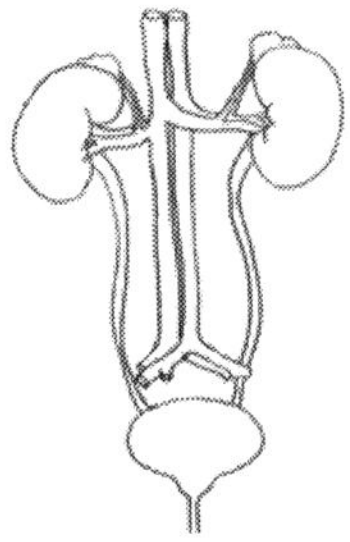

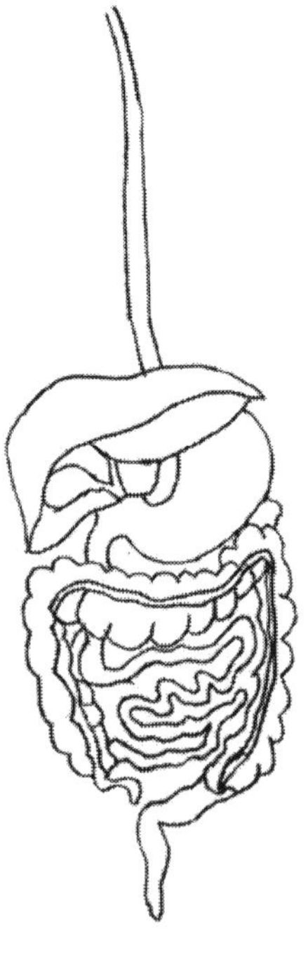

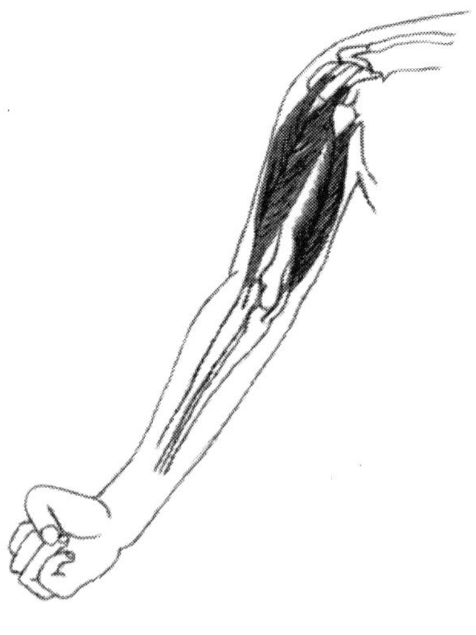

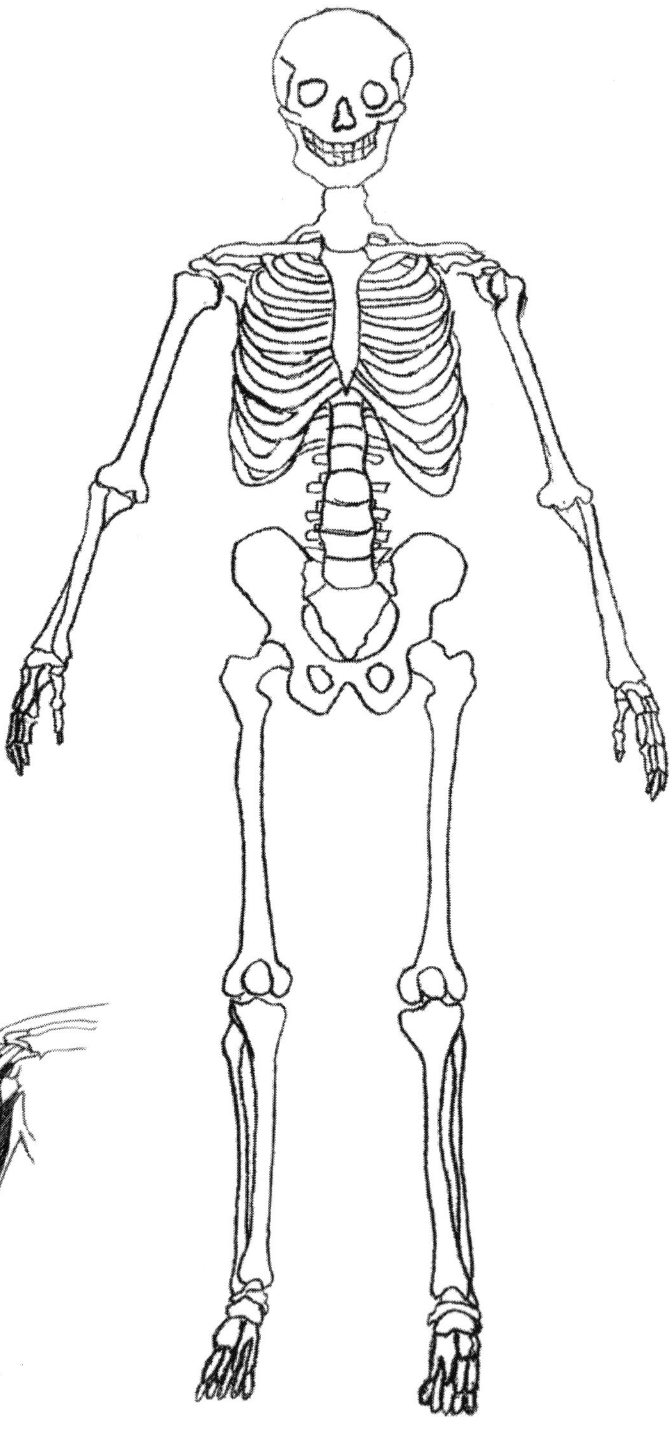

Human Body Front

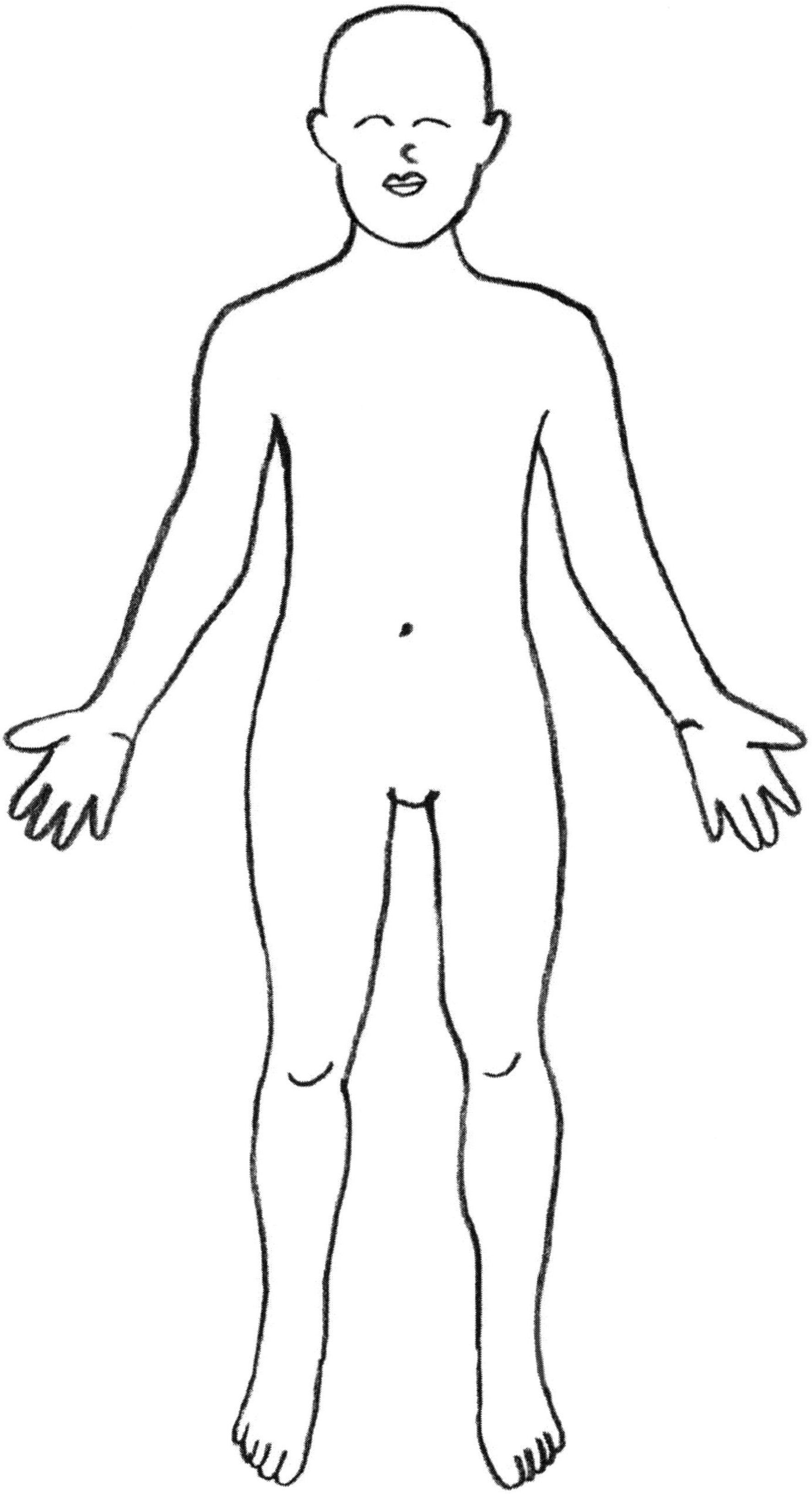

Human Body Back

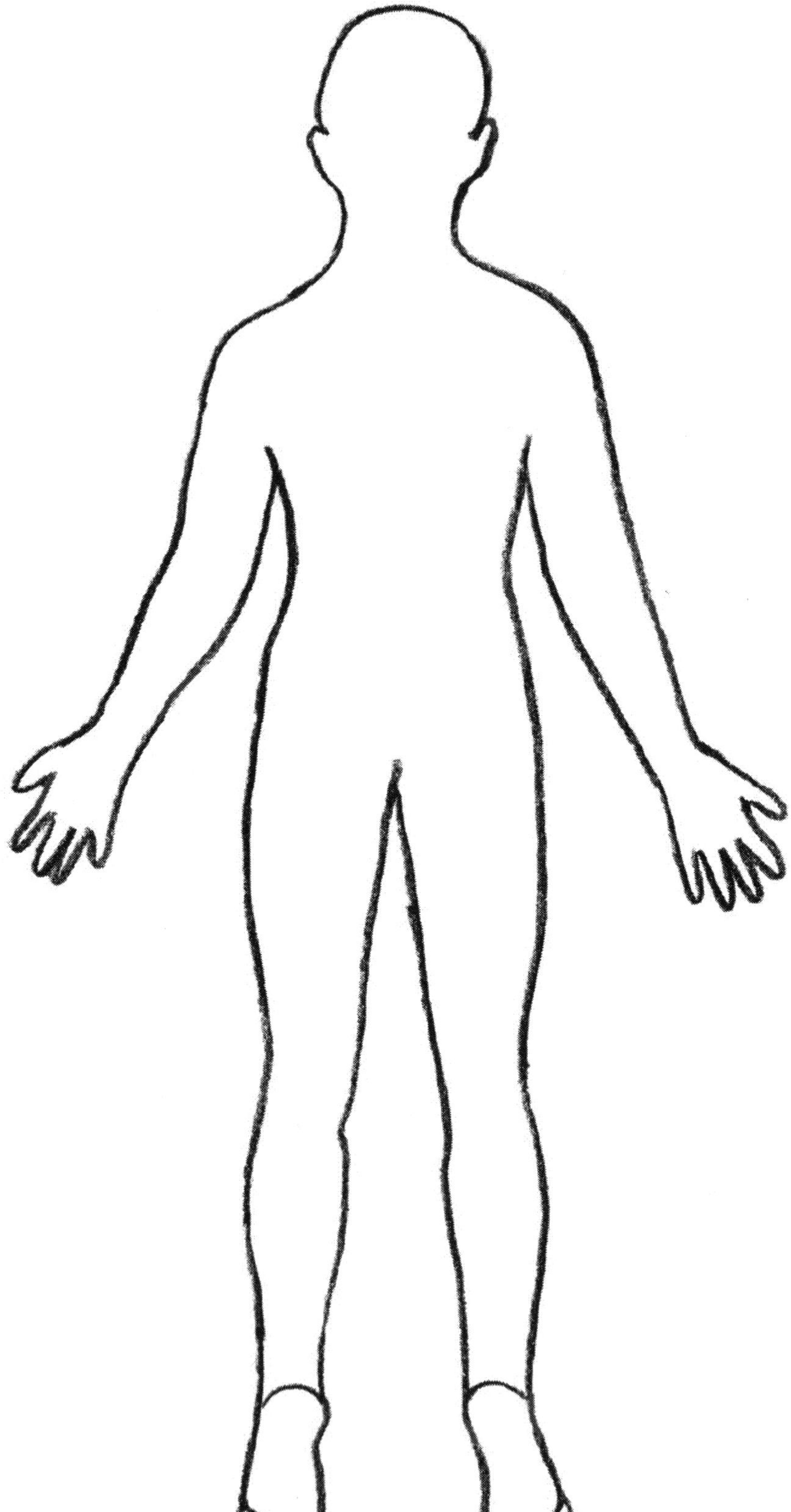

Rain Gauge Project

Week 1

Day	1	2	3	4	5	6	7
Predicted rainfall*							
Actual rainfall*							

*Don't forget to include units like inches (in) or centimeters (cm) with your measurements.

Week 2

Day	1	2	3	4	5	6	7
Predicted rainfall*							
Actual rainfall*							

*Don't forget to include units like inches (in) or centimeters (cm) with your measurements.

Rain Gauge Project

Week 3

Day	1	2	3	4	5	6	7
Predicted rainfall*							
Actual rainfall*							

*Don't forget to include units like inches (in) or centimeters (cm) with your measurements.

Week 4

Day	1	2	3	4	5	6	7
Predicted rainfall*							
Actual rainfall*							

*Don't forget to include units like inches (in) or centimeters (cm) with your measurements.

Week 5

Day	1	2	3	4	5	6	7
Predicted rainfall*							
Actual rainfall*							

*Don't forget to include units like inches (in) or centimeters (cm) with your measurements.

Week 6

Day	1	2	3	4	5	6	7
Predicted rainfall*							
Actual rainfall*							

*Don't forget to include units like inches (in) or centimeters (cm) with your measurements.

Constellation Page

Constellation Name ______________________________

What It Looked Like

What I Learned

Constellation Cards

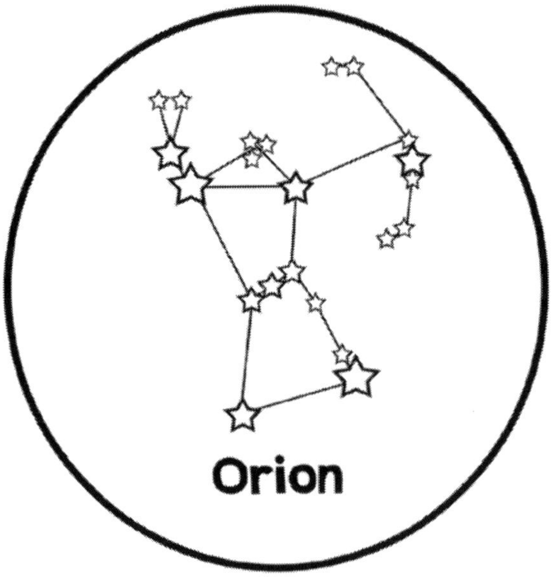

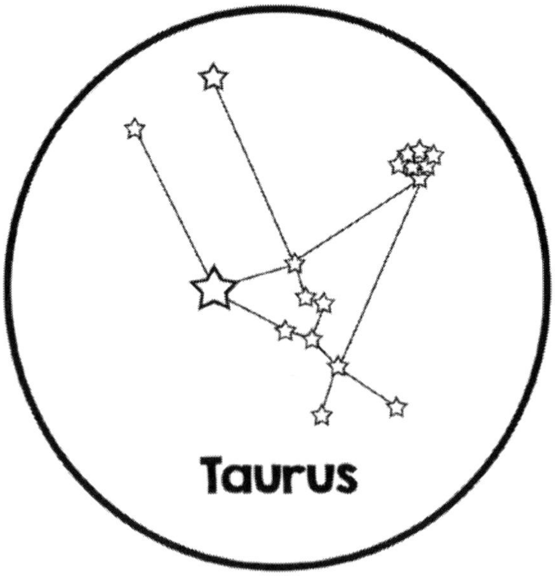

Constellation Cards

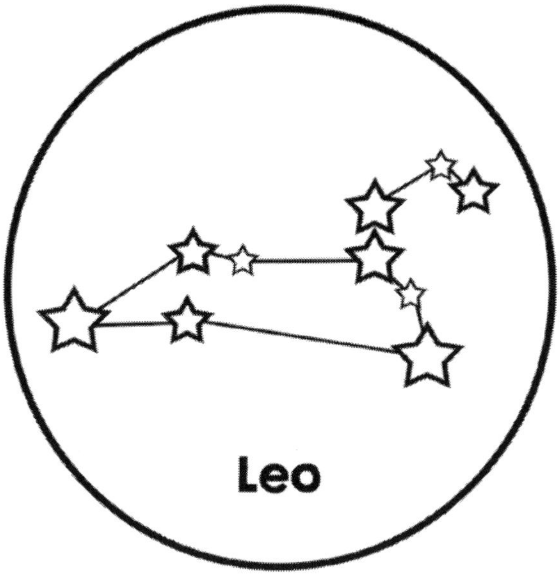

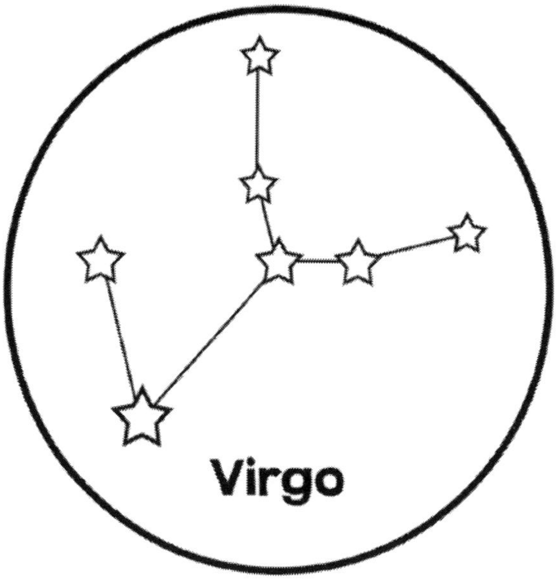

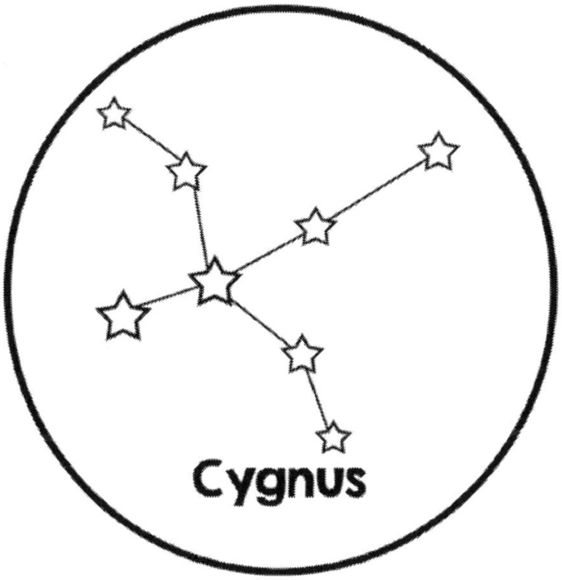

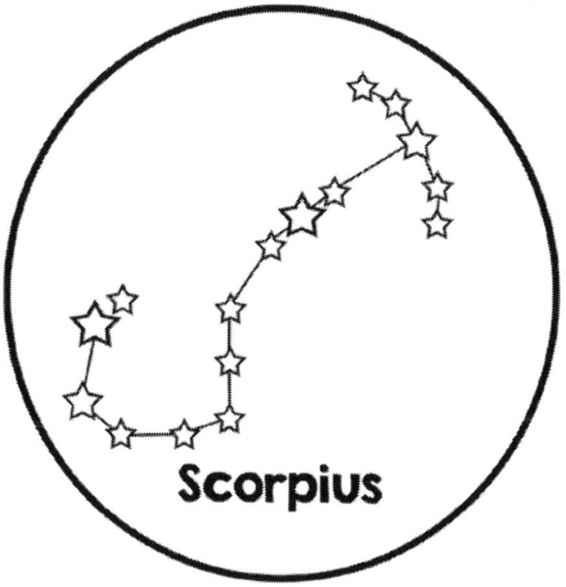

Constellation Cards

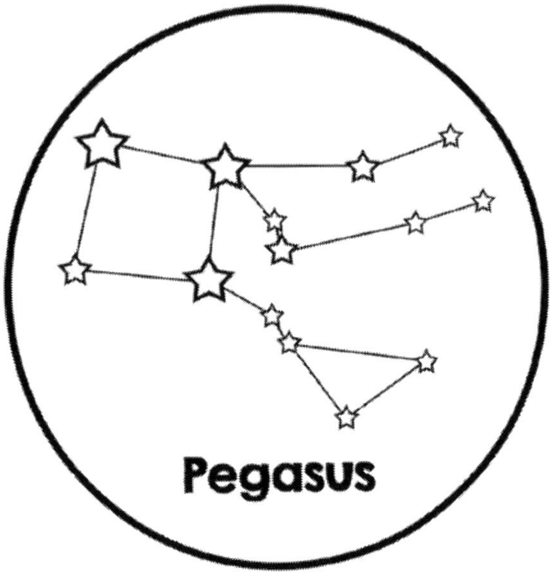

Neutralization

Neutralization is a chemical reaction that happens when an acid and a base cancel one another out. The end result in a neutral solution.

In water, acids like to donate hydrogen ions ($H+$). On the hand, bases in water like to donate hydroxide ions ($OH-$). The two ions are very attracted to each other. The hydrogen ion and the hydroxide ion come together to form something we are very familiar with – water!

There are other ions left in the solution after the acid donates its hydrogen ion and the base donates it hydroxide ion. These ions come together to form a salt.

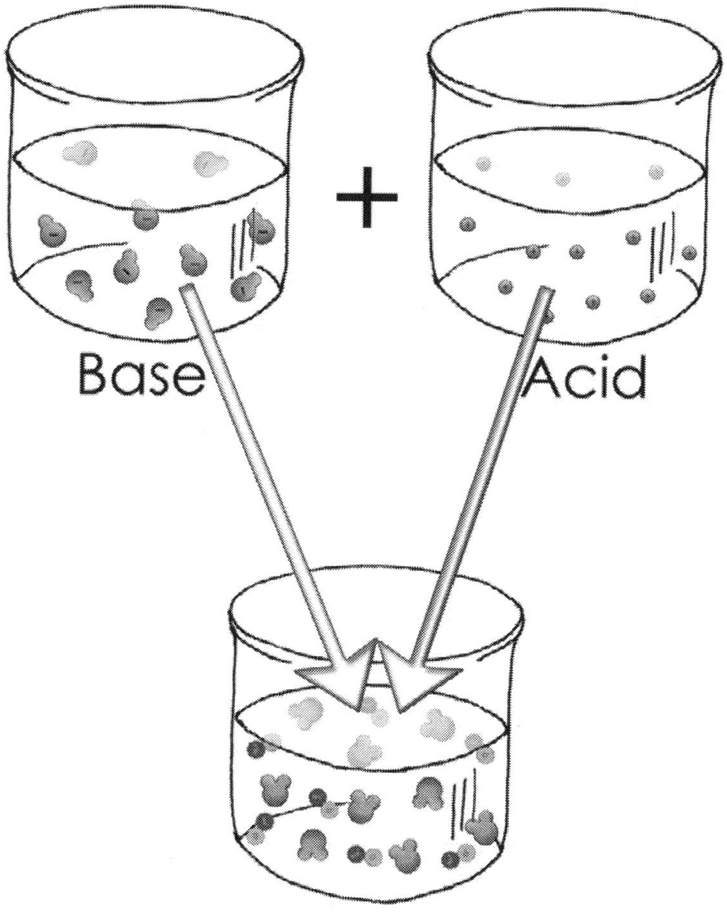

When you mix an acid and a base together in the right proportions, all of the ions will react and the solution will be come neutral.

SCIENCE CHUNKS

Glossary

Year B Glossary

A

- **Acid** – A chemical that dissolves in water and can neutralize a base. Weak acids taste sour.
- **Alveoli** – Tiny air bags found in your lungs.
- **Atmosphere** – A layer of gas that surrounds a planet.

B

- **Bacteria** – A group of microscopic organisms that can cause diseases.
- **Base** – A chemical that dissolves in water and can neutralize an acid. Weak bases taste bitter.
- **Blood vessel** – A tube that carries blood through the body.

C

- **Cells** – A tiny, living units that is the building blocks of all living things.
- **Climate** – The average weather in an area over a given period of time.
- **Conduction** – The transfer of heat through direct contact.
- **Constellation** – A group of stars that when viewed from Earth form the outline of an object or figure.
- **Convection** – The transfer of heat through the movement of a liquid and gas.

D

- **Diffusion** – The spreading out of a gas to fill the available space.
- **Digestion** – The process by which your food is broken down.

E

- **Energy** – The ability to do work.
- **Energy Chain** – A way of showing how energy changes into different forms.
- **Evaporation** – The process by which the surface molecules of a liquid escape into a vapor.

F

- **Fermentation** – A chemical reaction that breaks down sugar into carbon dioxide and alcohol.

G

H

- **Heat** – A form of energy that flows from one place to another because of differences in temperature.

I

- **Indicator** – A substance that changes color in the presence of an acid or base.

J

K

- **Kidney** – The organ in the body responsible for removing waste from itself and regulating the body's fluid levels.

L

M

- **Microbes** – A microscopic organism that can cause disease.
- **Muscle** – A type of tissue that makes the bones of your body move and that is controlled by your brain.

N

- **Natural cycle** – A naturally occurring event that happens over and over again in the same order.
- **Neuron** – A nerve cell that makes up the nervous system and carries electrical messages throughout the body.
- **Neutralization** – A reaction where one substance fully or partly cancels out another.
- **Nuclear Fission** – The splitting apart of atomic particles to create energy.
- **Nuclear Fusion** – The joining of atomic particles to create energy.

O

P

- **pH** – A scale from 0 to 14 used to measure the strength of acids and bases.
- **Physical Change** – A change that occurs in which no new substances are made.
- **Precipitation** – Rain, snow, sleet, or hail that falls to the ground.

Q

R

- **Radiation** – The transfer of heat through indirect contact.

S

- **Salt** – An ionic compound that when dissolved in water makes positive and negative.
- **Senses** – The ability of the body to take in and respond to information from its surroundings.
- **Skeleton** – The framework of 206 bones that supports your body; it allows you to move and protects certain organs.
- **Solar Energy** – Energy from the sun.
- **Star** – A huge ball of exploding gas.
- **States of Matter** – The different forms in which a substance can exist: solid, liquid, and gas.
- **Sublimation** – A change from solid to gas without going through liquid form.
- **Surface Tension** – A force that pulls together molecules on the surface of a liquid.

T

- **Temperature** – A measure of how much heat an object has.
- **Thunderstorm** – A storm with thunder and lightning.
- **Tornado** – A spinning funnel of wind.

U

V

- **Vapor** – Tiny droplets of water in the air.
- **Volume** – The space occupied by matter.

W

- **Wind Energy** – Energy from the wind.

X

Y

- **Yeast** – A microscopic, single-celled fungus. It is used in fermentation.

Z

SCIENCE CHUNKS

Review Sheets

Human Body Review Sheet

1. Your body has _____________ of cells.

a few   a couple hundred   billions

2. **True or False:** Genes help to determine what we will look like.

3. Circle the two layers of skin.

Dermis   Prodermis   Epidermis

4. **True or False:** Hair is alive and hurts when you cut it.

5. Circle all the things that your skeleton does.

Glows in the dark   Supports your body

Allows you to move   Protects certain organs

6. Your skull protects your _____________.

stomach   toes   brain   lungs

7. Muscles are controlled by your ______________.

mouth brain fingers

8. **True or False:** Joints help your body to move.

9. A ______________ is a nerve cell that carries electrical messages.

cell neuron muscle

10. **True or False:** The brain is not a part of the nervous system.

11. Match the body parts to the sense that uses it.

Nose	Sight
Ear	Hearing
Eye	Taste
Skin	Smell
Tongue	Feel

12. The heart ____________ blood through the body.

climbs pumps runs

13. The ____________ carry blood through the body.

blood vessels bones muscles

14. **True or False:** The lungs take in air.

15. The lungs give (oxygen / carbon dioxide) to the blood and remove (oxygen / carbon dioxide) from the blood.

16. Digestion is the process by which your food is ________________.

put back together broken down

17. __________________ help to chew up your food.

Your heart Your teeth Your eyes

18. Your kidneys get rid of ______________.

blood air waste

19. **True or False:** Genes help to determine what we will look like.

20. Germs are ________________.

microscopic big banana-sized

21. My body's defense system is called the _______________ system.

digestive circulatory immune

Weather Review Sheet

1. **True or False:** The Sun provides both light and heat for our planet.

2. The atmosphere helps protect us from the ____________ heat and light.

   Moon 's   Sun's   stars

3. Name the four seasons.

   F__________________   W__________________

   S__________________   S__________________

4. **True or False:** The climate of an area is the average weather over a period of time.

5. Circle the following factors that can affect the weather.

   Water   Air   Heat

6. **True or False:** Changes in the weather occur on the surface of the Earth.

7. **True or False:** Thunderstorms have both thunder and lightening.

8. **True or False:** Rain is not a form of precipitation.

9. Match the correct step of the water cycle with the numbers in the image below.

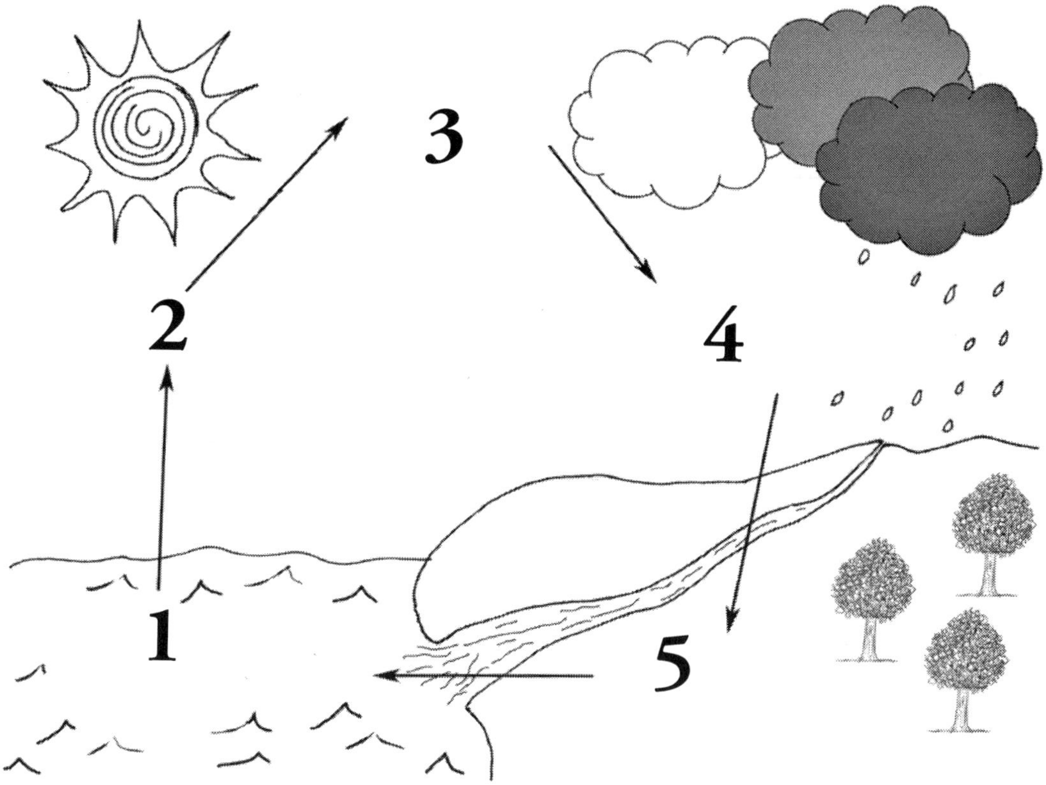

A. As the water vapor cools, the clouds become heavy and water falls as rain or snow.
B. Water vapor condenses to form clouds.
C. The sun's heat evaporates water from the world's oceans, lakes, and rivers.
D. The water rises into the air as water vapor.
E. Rainwater and snow melt flow back into the rivers, lakes, and oceans.

10. Match the following.

Flood	A long period of dry conditions and little water.
Drought	An overflow of a large amount of water.

11. **True or False:** Scientist, called meteorologists, can predict the weather.

Stars Review Sheet

1. Match the constellation picture with its name.

A.

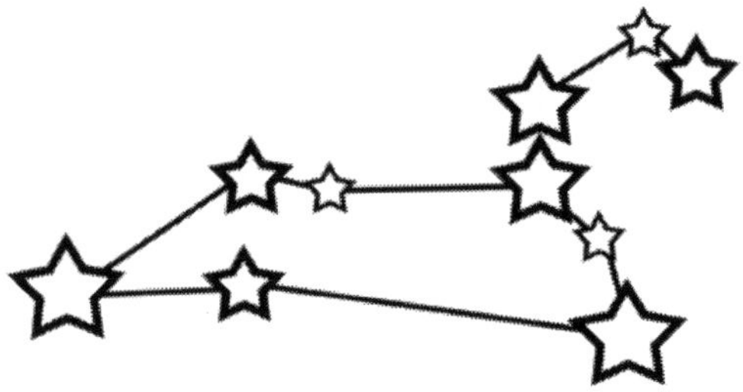

_____ Ursa Major

B.

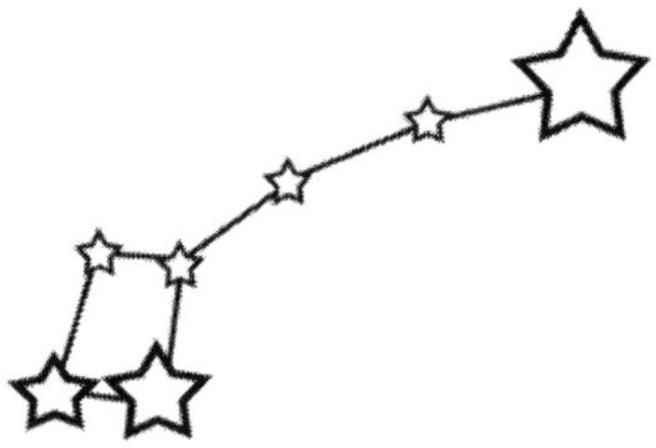

_____ Ursa Minor

C.

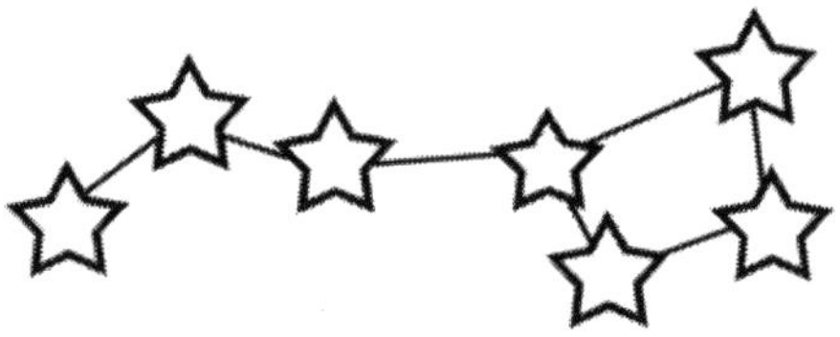

_____ Orion

D.

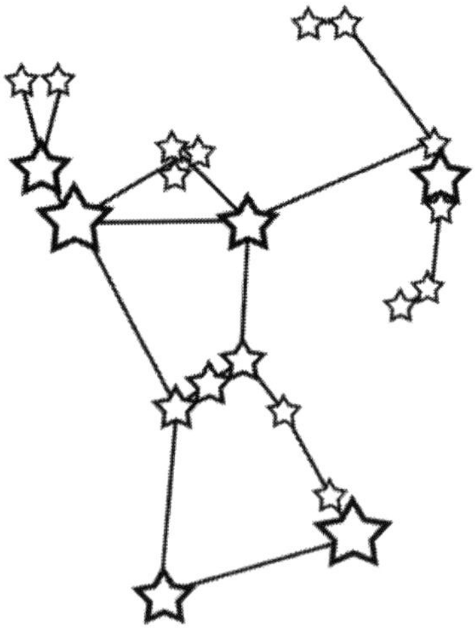

_____ Leo

2. A star is really a huge ball of exploding __________.

gas water air

3. Match the constellation picture with its name.

_____ Pegasus

_____ Cygnus

_____ Scorpius

_____ Andromeda

A.

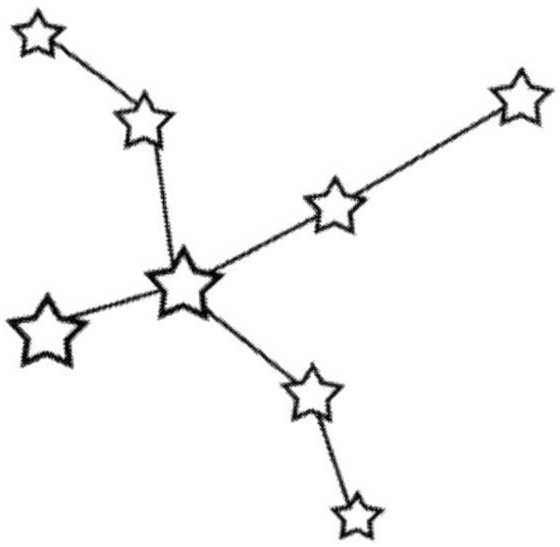

B.

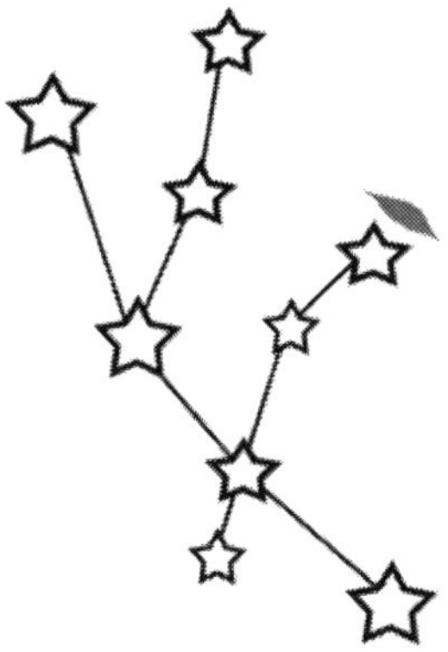

C.

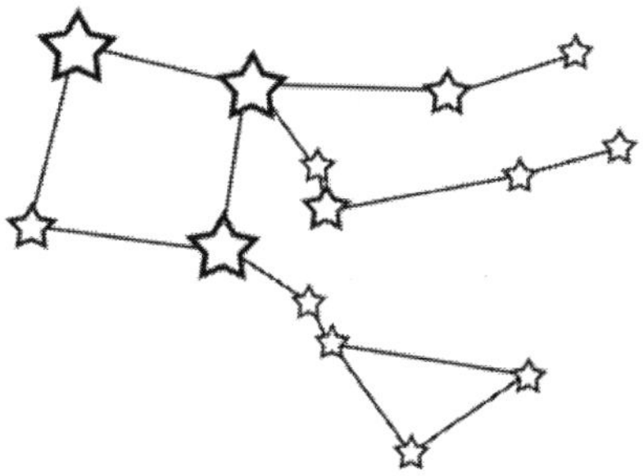

D.

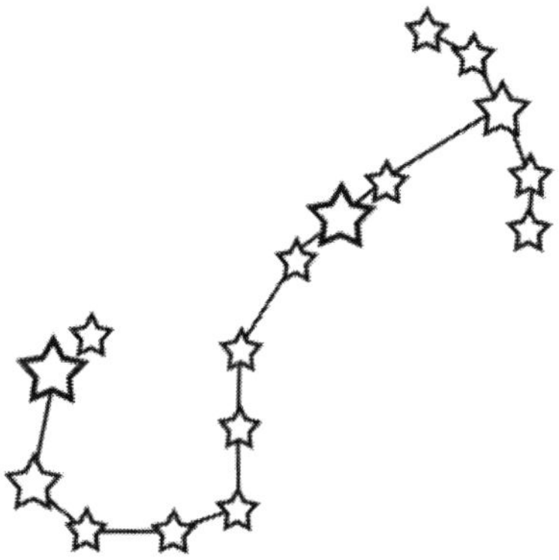

4. A group of stars that when viewed from space makes the outline of an object or figure is called a ________________________.

Matter Review Sheet

1. Match the state of matter.

Solid _____ A. Has no fixed shape or volume.

Liquid _____ B. Has a fixed shape and volume.

Gas _____ C. Has a fixed volume, but not a fixed shape.

2. **True or False:** Volume is the amount of space occupied by matter.

3. When a solid ____________________, it turns into a liquid.

When a liquid ____________________, it turns into a gas.

melts boils condenses

4. When a liquid ____________________, it turns into a solid.

When a liquid ____________________, it turns into a gas.

condenses freezes evaporates

5. Liquids can change ____________________, but the ____________________ remains the same.

volume   shape   color

6. **True or False:** Surface tension is the result of the molecules in a liquid being attracted to each other.

7. In diffusion, molecules move from an area of (high / low) concentration to an area of (high / low) concentration.

8. ____________________ is the push that gas molecules exert on a container.

volume   pressure   temperature

Acids and Bases Review Sheet

1. **True or False:** An acid dissolves in water and can taste sour.

2. **True or False:** A base dissolves in water and can taste sour.

3. pH stands for ____________________________.

potential of hydrology   power of hydrogren

4. Label the following pH's with what they represent - acid, base, or neutral.

pH 2 ____________________________

pH 7 ____________________________

pH 10 ___________________________

5. When you mix an acid and a base, you get ______________________________ and

a ___________________________.

6. A salt dissolves in water to form ____________________________.

bases   ions   acids

Pasteur Review Sheet

1. Match the following words to their definitions.

_____ **Fermentation**	A. A microscopic, single-celled fungus. It is used in fermentation.
_____ **Microbes**	B. A microscopic organism that can cause disease.
_____ **Yeast**	C. A chemical reaction that breaks down sugar into carbon dioxide and an alcohol.

2. **True or False:** Louis Pasteur was a teacher.

3. **True or False:** Louis Pasteur only tested his findings once before sharing them.

4. **True or False:** Pasteur concluded that the yeasts in the good batch were alive, feeding on the beet juice to produce alcohol. In the foul batch sample, he concluded that the black rods had taken over and stopped the yeast from making alcohol.

5. **True or False:** Pasteur found that microbes were in the air on the dust particles that floated around.

6. Why do you think someone should learn about Louis Pasteur?

____

____

____

____

____

Energy Review Sheet

1. Match the following types of energy.

Potential Energy	Energy of motion
Kinetic Energy	Energy that is stored

2. The Law of Conservation of energy says that ____________________ can neither be ____________________ or ________________________.

3. What is a renewable source of energy?

____

____

____

4. What is a nonrenewable source of energy?

____

____

____

5. **True or False:** Nuclear reactors use controlled fusion reactions.

6. **True or False:** In a nuclear reactor, the energy released heats up water to generate steam that turns a turbine and generates energy.

7. What is heat?

____

____

____

8. A ____________________ is an instrument that measures temperature.

9. Match the object with the way light passes through it.

____ Transparent object	A. Light does not pass through.
____ Translucent object	B. Light passes completely through.
____ Opaque object	C. Light passes partially through.

10. **True or False:** Different colors of light have the same wavelength.

Made in United States
Orlando, FL
10 November 2026

90506337R00077